FRANCIS GALTON AND ALFRED BINET: CORRESPONDENCE

Francis Galton & Alfred Binet

Correspondence

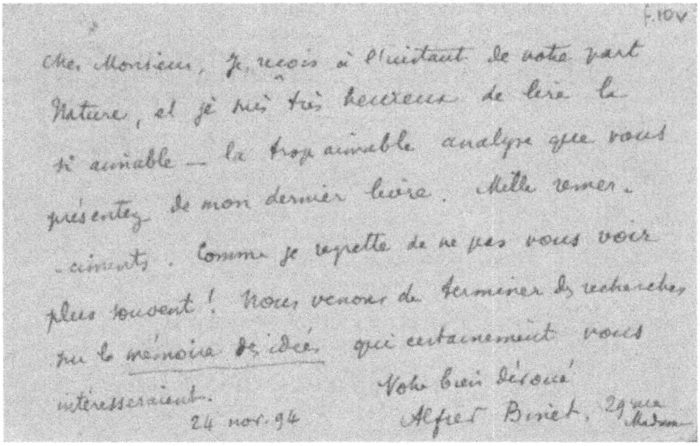
f.10v

Cher Monsieur, Je reçois à l'instant de votre part Nature, et je suis très heureux de lire la si aimable — la trop aimable analyse que vous présentez de mon dernier livre. Mille remerciements. Comme je regrette de ne pas vous voir plus souvent ! Nous venons de terminer des recherches sur la mémoire des idées qui certainement vous intéresseraient.

Votre bien dévoué
Alfred Binet, 29 rue Madame
24 nov. 94

Edited by Gavan Tredoux

galton.org

Francis Galton: a Lifetime of Exploration

1. Francis Galton's Nature and Nurture: 1822–1865.
2. Francis Galton's Genius: 1865–1911.

Supplements:

1. Sir Francis Sacheverel Darwin.
2. Francis Galton at Cambridge: Letters and Diaries, 1840–1844.
3. The Diary of Charles John Andersson: 1850–1851.
4. Francis Galton on Mars: the Discontinuous Variation Notebook.
5. Francis Galton and Alphonse de Candolle: Notes and Correspondence.
6. Francis Galton and Alfred Binet: Correspondence.

7. Francis Galton: Selected Papers, Notebooks and Diaries.

All content provided in this series is for educational purposes only. Any proceeds go to maintaining the non-profit `http://galton.org`.

Contents

1 Introduction **1**

2 Correspondence **17**

2.1 1891-11-25 Binet to FG . . . 17

2.2 1892-03-19 Binet to FG . . . 23

2.3 1893-02-10 Binet to FG . . . 28

2.4 1893-02-12 Binet to FG . . . 31

2.5 1893-04-20 Binet to FG . . . 35

2.6 1893-05-22 Binet to FG . . . 41

2.7 1894-08-22 Binet to FG . . . 49

2.8 1894-10-11 Binet to FG . . . 55

2.9 1894-11-24 Binet to FG . . . 59

3 Galton on Binet, 1894 **67**

4 Galton's Notes on Inaudi **83**

5 Charcot's Report on Inaudi **101**

List of Figures **117**

Bibliography **121**

Register **123**

Chapter 1

Introduction

This brief volume contains previously unpublished correspondence between the polymath Sir Francis Galton (1822–1911) and the French psychologist Alfred Binet (1857–1911). Both men were pioneers in the field of mental testing.

Among his many other accomplishments, Galton devised physical reaction-time tests for intelligence—see *Francis Galton: a Lifetime of Exploration*[1] and `http://ga`

[1] Gavan Tredoux (2023a). *Francis Galton's Genius. 1865–1911.* galton.org, 2023; Gavan Tre-

`lton.org` for full details on Galton.

Binet first devised the paper-and-pencil intelligence tests that we are now familiar with, for use with schoolchildren. (Later these would be called 'IQ' tests when expressed as quotients of mental age but that was not Binet's idea.) These tests combine items that have modest to low individual item reliability into a composite test with high overall reliability. They were rapidly adopted in the UK (by Cyril Burt) and the USA (by H. H. Goddard and Lewis Terman), after being translated into English.[2]

The two scientists communicated from the early 1890s onward—the correspondence was initiated by Binet—sharing interests in composite photographs, calculating prodigies, mental fatigue and imagery. The letters suggest that Binet and Galton met in person at least once, but

---

doux (2023b). *Francis Galton's Nature and Nurture. 1822–1865.* galton.org, 2023.

2 Burt 1952; Haier 2017; Jensen 1980, 1998.

no other details are currently known. In fact Galton had been reading Binet since the mid-1880s, recommending the Frenchman's book on *The Psychology of Reasoning* to Max Müller while exchanging notes with the philologist about 'Thoughts without Words'.[3] Müller ordered it.

Galton reviewed Binet's 1894 book on the *Psychology of mental arithmeticians and blindfold chess players* for *Nature*.[4] On Binet's recommendation, one of the most famous of these calculating prodigies, the Piedmontese Jacques (Giacomo) Inaudi (1867–1950), was presented to Galton for detailed examination when the boy toured

---

3 Alfred Binet (1886). *La psychologie du raisonnement, recherches expérimentales par l'hypnotisme.* Paris: Alcan, 1886; Francis Galton (1887b). "Thoughts without Words [Letters]". In: *Open Court* 1 (1887), pp. 441–4, 472–4, 498–50.

4 Francis Galton (1894b). "Psychology of mental arithmeticians and blindfold chess-players". In: *Nature* 51 (Nov. 22, 1894). Review of Alfred Binet, Psychologie des Grands Calculateurs et Joueurs, pp. 73–4.

England the previous year.[5]

In 1886 Galton had visited an 'Asylum for Idiots' at Earlswood, where he had toured a few years earlier with George Romanes, and another at Darenth, near Dartford. 'Span of prehension' (now called digit span) tests were given to a handful of ordinary female inmates, who were asked to recall strings of random numerals in the order read to them. Predictably they did rather poorly, though better at Darenth, who were 'merely imbecile'. But three male inmates at Earlswood had remarkable memories for historical dates, which they enjoyed memorizing from books, even though their digit span was no better than others.[6] These cases were unlike the otherwise-ordinary French calculating sensation Jacques Inaudi, who had been marvelling

[5] Gavan Tredoux (2023a). *Francis Galton's Genius. 1865–1911.* galton.org, 2023.

[6] Francis Galton (1887a). "Supplementary notes on 'Prehension in idiots'". In: *Mind* 12 (Jan. 1887), pp. 79–82.

audiences since 1880, to the extent that he was formally investigated by a committee of French scientists led by Charcot in 1892.

The following year Inaudi was publicly exhibited in England. His handler bore a letter of introduction from 'Professor' Binet to Galton, and the French committee's report.[7] After discreetly assembling a local committee at a Royal Society *conversazione*, assuring the strictest confidence for all, Galton received Inaudi and his handler at Rutland Gate on the morning of May 11, for a *seance*. The scientists examined the boy's mental powers and found that he had a 'prodigious memory for figures', that he could easily perform feats like calculating cube roots of 7 figure numbers in his mind, but (somewhat obtusely) that his arithmetical abilities did not exceed those of an able bank clerk

---

7 Galton's notes are wrong here, as Binet was not then a professor at any institution.

> once the time saved writing things down was accounted for. In his mental operations he did not seem to operate visually, relying rather on sounds, with a special dislike for the number 7. 'He had that wearied look in the eyes & brow that I have seen in other boy calculators', Galton noted. In 1894 Binet published *Psychologie des grands calculateurs et joueurs d'échecs* (Psychology of mental arithmeticians and blindfold chess-players), which covered the non-visualiser Inaudi and a Greek calculating prodigy who did visualise, Diamandi. Galton reviewed it favourably in *Nature*.[8]

Only Binet's letters to Galton have been traced. Galton's own letters to Binet have not been traced, and may have perished in the general destruction and dispersal of Binet's papers after his death. According to Alexandre Klein, such papers of Binet as remain have had to be reconstituted from

[8] See chapter 3, p. 67.

miscellaneous sources all over Europe, and from previously unknown remains possessed by his descendants. There is now a substantial collection, the "Binet Family Fund", in the National Library of France, containing the papers that his daughters had kept until their deaths.[9]

The letters reproduced here were transcribed to text from the handwritten French MSS by Odile Thomas. They were translated by the editor, who has tried to present an idiomatic translation rather than a strictly literal one. Each letter is presented with the text of the French original for comparison, and images of the MSS. They are supplemented by Galton's review of Binet, and his notes on the inspection of Jacques Inaudi.

---

9 Klein 2016, 2018.

Please report any errors to the editor via `galton.org`.

Gavan Tredoux, August 2022.

Figure 1.1: Alfred Binet.

Figure 1.2: Francis Galton in the 1890s.

Figure 1.3: Jacques (Giacomo) Inaudi as a boy.

Figure 1.4: Binet with promoter 'M. de Thorcey'.

Figure 1.5: Francis Galton and his home at 42 Rutland Gate, Kensington.

Figure 1.6: Francis Galton in 1904 at age 82.

Figure 1.7: Francis Galton in 1904 at age 82.

Figure 1.8: The Binet Family.

Chapter 2

Correspondence

2.1 1891-11-25: Alfred Binet to Francis Galton

Physiological Psychology Laboratory, Sorbonne

Paris, 25 Nov. 91

Sir

I venture to make a request to you on behalf of the [Physiological Psychology] laboratory, the management of which M. Beaunis[1] was kind enough to entrust to me for the winter. Do you still have some composite photographs, and could you donate a copy to the laboratory? That copy would be received with the greatest appreciation.

Please accept, Sir, my warmest approbations and best regards.

Alfred Binet

Paris, 29 rue de Madame

---

[1] Henri-Étienne Beaunis (1830–1921), psychologist.

Physiological Psychology Laboratory, Sorbonne

Paris, 25 nov. 91

Monsieur

Je me permets de venir vous faire une demande dans l'intérêt du laboratoire, dont M Beaunis a bien voulu me confier la direction pendant l'hiver. Possédez-vous encore quelques photographies composites, et pourriez vous en donner un exemplaire au laboratoire? Cet exemplaire serait reçu avec la plus grande reconnaissance.

Veuillez agréer, Monsieur, l'hommage de ma haute considération et de ma particulière estime.

Alfred Binet

Paris, 29 rue de Madame

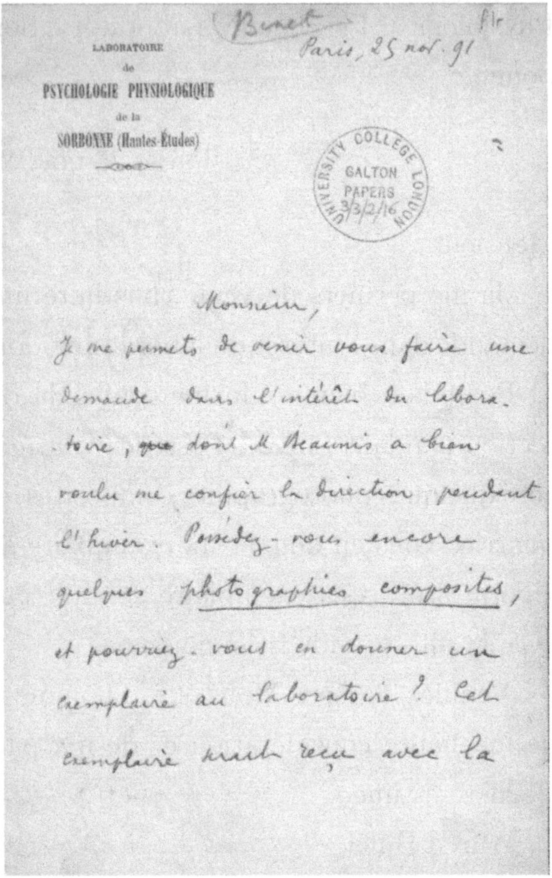

Binet

F1r

LABORATOIRE
de
PSYCHOLOGIE PHYSIOLOGIQUE
de la
SORBONNE (Hautes-Études)

Paris, 25 nov. 91

Monsieur,

Je me permets de venir vous faire une demande dans l'intérêt du labora-toire, ~~que~~ dont M Beaunis a bien voulu me confier la direction pendant l'hiver. Possédez-vous encore quelques <u>photographies composites</u>, et pourriez-vous en donner un exemplaire au laboratoire? Cet exemplaire serait reçu avec la

Figure 2.1: Binet to Galton 1891-11-25 Page 1

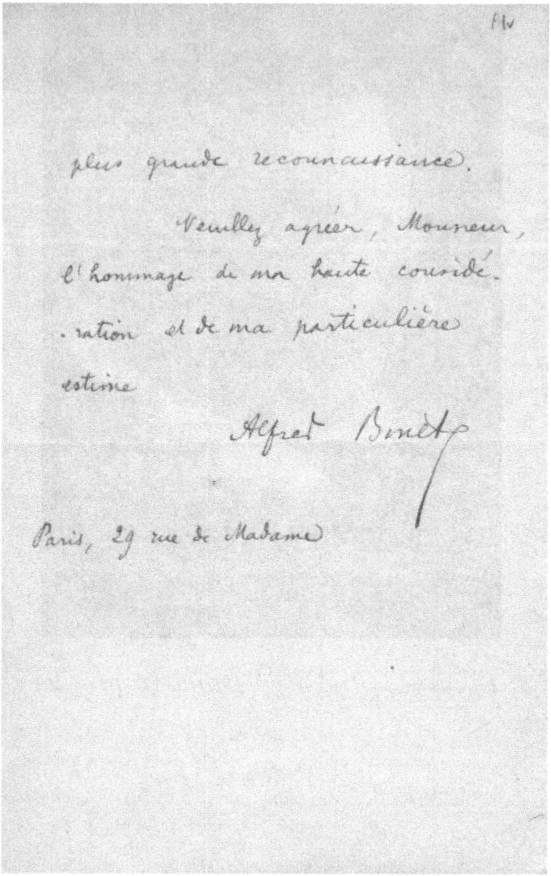

plus grande reconnaissance.

Veuillez agréer, Monsieur,
l'hommage de ma haute considé-
-ration et de ma particulière
estime

Alfred Binet

Paris, 29 rue de Madame

Figure 2.2: Binet to Galton 1891-11-25 Page 2

Figure 2.3: Alfred Binet in profile

2.2 1892-03-19: Alfred Binet to Francis Galton

29 rue de Madame
19 March 92

Dear Sir

Thank you for the kind letter you wrote to me. I am very happy that my little volume on reasoning[2] meets with your approval — it would be very useful for us to keep in touch with you, our laboratory has much to gain from it, for I know that you have devised many ingenious methods of studying the sound man from the psychological point of view. Would it be too much to ask you for your photograph for our laboratory collection?

Please accept, dear Sir, the expression of my respectful and devoted feelings,

---

[2] Alfred Binet (1886). *La psychologie du raisonnement, recherches expérimentales par l'hypnotisme.* Paris: Alcan, 1886.

Alfred Binet

29 rue de Madame
19 mars 92

Cher Monsieur

Je vous remercie de la lettre si aimable que vous m'avez écrite. Je suis bien heureux que mon petit volume sur le raisonnement ait eu votre approbation – il serait très utile pour nous de rester en relation avec vous, notre laboratoire a beaucoup à y gagner, car je sais que vous avez imaginé beaucoup de méthodes ingénieuses pour étudier l'homme sain au point de vue psychologique. Serait-il indiscret de vous demander votre photographie pour notre collection du laboratoire?

Veuillez agréer, cher Monsieur, l'expression de mes sentiments respectueux et dévoués,

Alfred Binet

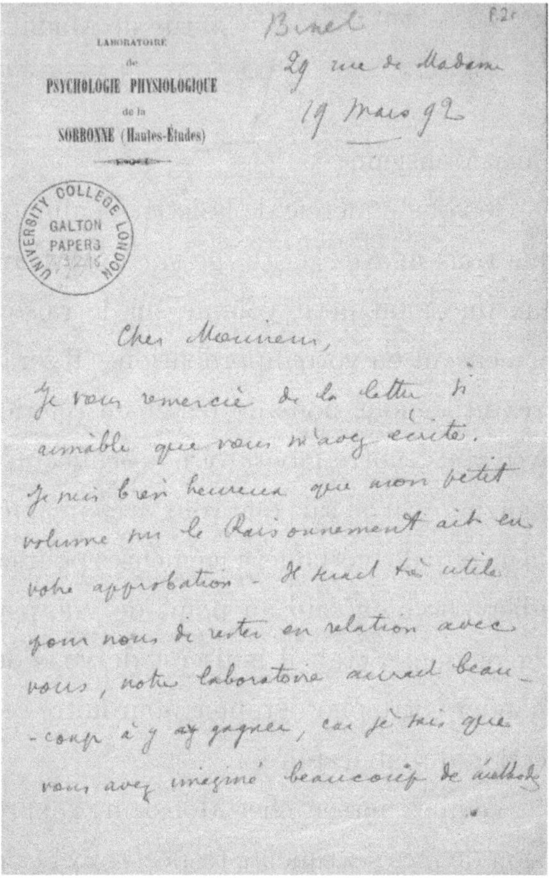

LABORATOIRE
de
PSYCHOLOGIE PHYSIOLOGIQUE
de la
SORBONNE (Hautes-Études)

Binet
29 rue de Madame
19 mars 92

Cher Monsieur,

Je vous remercie de la lettre si aimable que vous m'avez écrite. Je suis bien heureux que mon petit volume sur le Raisonnement ait eu votre approbation. Il serait très utile pour nous de rester en relation avec vous, notre laboratoire aurait beaucoup à y gagner, car je sais que vous avez imaginé beaucoup de méthodes

Figure 2.4: Binet to Galton 1892-03-19 Page 1

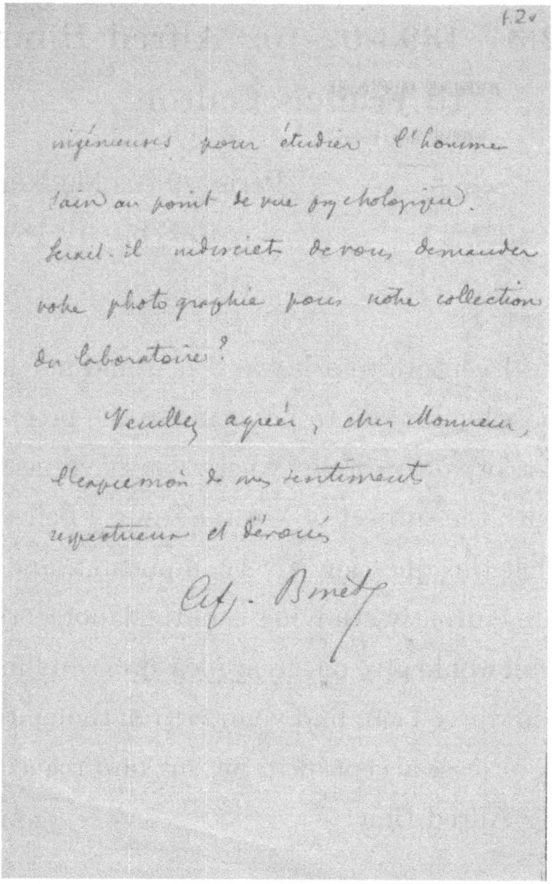

f.2v

ingénieuses pour étudier l'homme
sain au point de vue psychologique.
Serait-il indiscret de vous demander
votre photographie pour notre collection
du laboratoire?

Veuillez agréer, cher Monsieur,
l'expression de mes sentiments
respectueux et dévoués

Alf. Binet

Figure 2.5: Binet to Galton 1892-03-19 Page 2

2.3 1893-02-10: Alfred Binet to Francis Galton

Paris, 29 rue Madame
10 February 1893

Dear Sir

I venture to ask you if it would not be possible for you to point me to the precise list of your articles where you have dealt with the subject of Mental Tests. I believe that this question is very important, and I am currently studying it in the laboratory. You would give me great pleasure by telling me where I can find your written thoughts.

Please accept, dear Sir, my best regards,

Alfred Binet

Paris, 29 rue Madame
10 fevrier 1893

Cher Monsieur

Je viens vous demander s'il ne vous serait pas possible de m'indiquer exactement la liste de vos articles où vous avez traité la question des Mental Tests. Je crois que cette question est très importante, et je l'étudie en ce moment au laboratoire. Vous me feriez grand plaisir en m'indiquant par conséquent où je pourrai trouver votre pensée écrite.

Veuillez agréer, cher Monsieur, mes meilleurs sentiments,

Alfred Binet

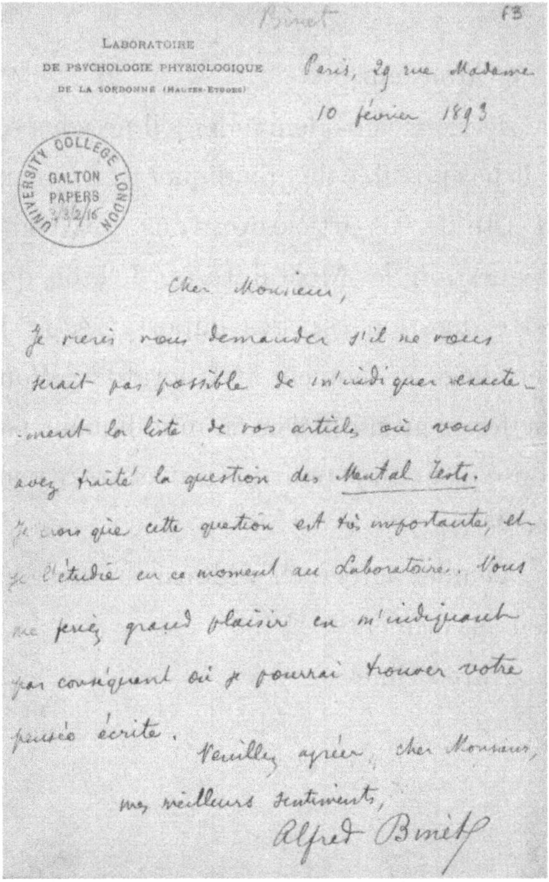

Binet F3

LABORATOIRE
DE PSYCHOLOGIE PHYSIOLOGIQUE
DE LA SORBONNE (HAUTES-ÉTUDES)

Paris, 29 rue Madame
10 février 1893

Cher Monsieur,

Je viens vous demander s'il ne vous serait pas possible de m'indiquer exactement la liste de vos articles où vous avez traité la question des Mental Tests. Je crois que cette question est très importante, et je l'étudie en ce moment au Laboratoire. Vous me feriez grand plaisir en m'indiquant par conséquent où je pourrai trouver votre pensée écrite.

Veuillez agréer, cher Monsieur, mes meilleurs sentiments,

Alfred Binet

Figure 2.6: Binet to Galton 1893-02-10

2.4 1893-02-12: Alfred Binet to Francis Galton

Paris, 29 rue de Madame
February 12, 1893

Dear Sir

I hasten to reply to your kind letter. Don't send me your article on mental fatigue,[3] I have it—Féré's address is:[4]

37 Boulevard Saint-Michel
Paris

I am going to read with great pleasure

[3] Francis Galton (1888). "Remarks on Replies by Teachers to Questions Respecting Mental Fatigue". In: *Journal of the Anthropological Institute* 18 (Apr. 10, 1888), pp. 157–68.

[4] Charles Samson Féré (1852–1907) who had like Binet been an assistant to the neurologist Dr. Jean-Martin Charcot (1825–1893). He was then chief medical officer at the Hospice Bicêtre. Galton wrote to Féré, who replied promptly on 1893-02-16, see GALTON/3/3/6/4.

the articles that you point me to in *Nature*.[5]

Please accept, dear Sir, the expression of my very devoted feelings,

Alfred Binet

[5] It is not clear which articles Binet was referring to.

Paris, 29 rue de Madame
12 fevrier 1893

Cher Monsieur

Je me hâte de répondre à votre aimable lettre. Ne m'envoyez pas votre article sur la fatigue mentale, je l'ai – l'adresse de Féré est:

37 Boulevard St Michel
Paris

Je vais lire avec grand joui les articles que vous m'indiquez dans Nature.

Veuillez recevoir, cher Monsieur, l'expression de mes sentiments bien dévoués,

Alfred Binet

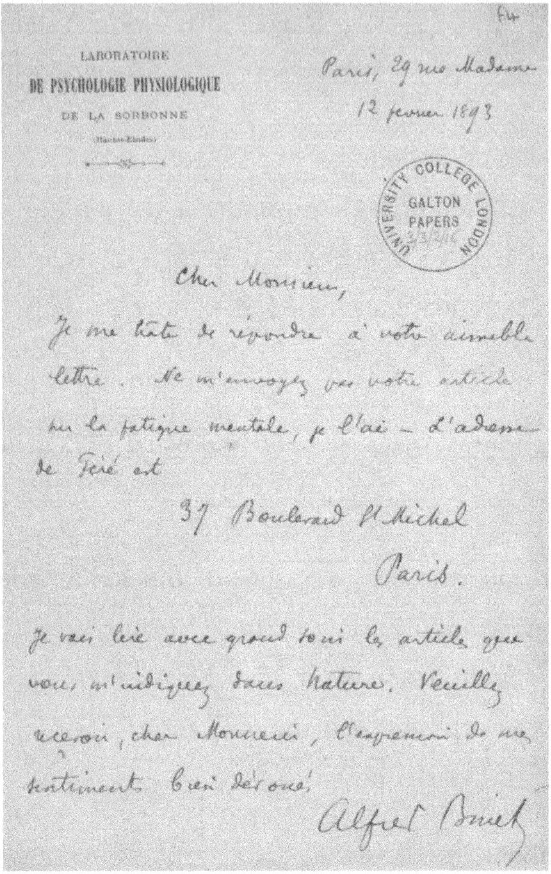

LABORATOIRE
DE PSYCHOLOGIE PHYSIOLOGIQUE
DE LA SORBONNE
(Hautes-Études)

Paris, 29 rue Madame
12 février 1893

Cher Monsieur,

Je me hâte de répondre à votre aimable lettre. Ne m'envoyez pas votre article sur la fatigue mentale, je l'ai — L'adresse de Féré est

37 Boulevard St Michel
Paris.

Je vais lire avec grand soin les articles que vous m'indiquez dans Nature. Veuillez recevoir, cher Monsieur, l'expression de mes sentiments bien dévoués,

Alfred Binet

Figure 2.7: Binet to Galton 1893-02-12

2.5 1893-04-20: Alfred Binet to Francis Galton

Paris, 29 rue Madame
April 20, 1893

Dear Sir

I recommend to your scientific curiosity, as well as to all your good will, Mr. Jacques Inaudi, a young calculating prodigy, on whom I published a study about a year ago. M. Jacques Inaudi, who is indeed of the family of Mondeux, Bidder and Colburn,[6] shows a novel psychological character; in his operations of mental calculation, he does not see the figures, he hears them and pronounces them. I have no doubt that this question interests you, you who have done everything for the study of mental imagery.

---

[6] Henry Mondeaux (1826–1861), George Parker Bidder (1806–1878) and Zerah Colburn (1804–1839) were calculating prodigies. By 'family' Binet means type.

Please accept, dear Sir,—with my thanks for the sending of your interesting brochure,—the expression of my best regards,

Alfred Binet

Paris, 29 rue Madame
20 avril 1893

Cher Monsieur

Je recommande à votre curiosité scientifique, ainsi qu'à toute votre bienveillance, M. Jacques Inaudi, un jeune calculateur prodige, sur lequel j'ai publié une étude il y a environ un an. M. Jacques Inaudi, qui est bien de la famille de Mondeux, des Bidder et des Colburn, présente un caractère psychologique original; dans ses opérations de calcul mental, il ne voit pas les chiffres, il les entend et les prononce. Je ne doute pas que cette question vous intéresse, vous qui avez tout fait pour l'étude des images mentales.

Veuillez agréer, cher Monsieur, – avec mes remerciements pour l'envoi de votre intéressante brochure, – l'expression de mes sentiments les plus distingués,

Alfred Binet

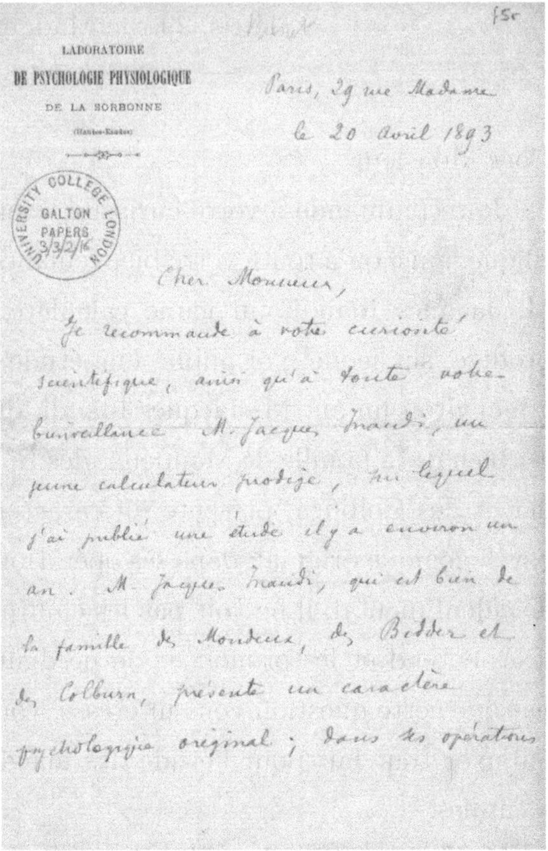

Binet

15r

LABORATOIRE
DE PSYCHOLOGIE PHYSIOLOGIQUE
DE LA SORBONNE
(Hautes-Études)

Paris, 29 rue Madame
le 20 Avril 1893

Cher Monsieur,

Je recommande à votre curiosité scientifique, ainsi qu'à toute votre bienveillance M. Jacques Inaudi, un jeune calculateur prodige, sur lequel j'ai publié une étude il y a environ un an. M. Jacques Inaudi, qui est bien de la famille de Mondeux, de Bidder et de Colburn, présente un caractère psychologique original ; dans ses opérations

Figure 2.8: Binet to Galton 1893-04-20 Page 1

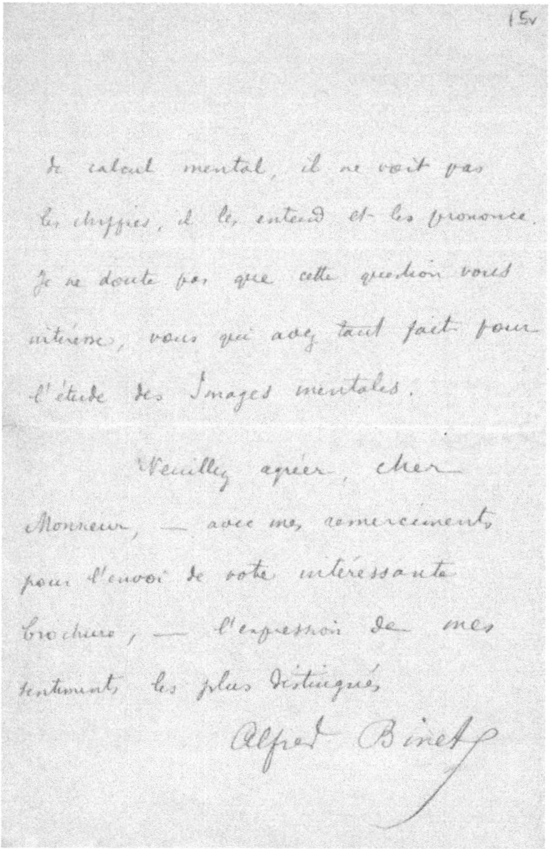

15v

de calcul mental, il ne voit pas les chiffres, il les entend et les prononce. Je ne doute pas que cette question vous intéresse, vous qui avez tant fait pour l'étude des Images mentales.

Veuillez agréer, cher Monsieur, — avec mes remerciements pour l'envoi de votre intéressante brochure, — l'expression de mes sentiments les plus distingués,

Alfred Binet

Figure 2.9: Binet to Galton 1893-04-20 Page 2

2.6 1893-05-22: Alfred Binet to Francis Galton

Paris, 29 rue Madame
22 May 1893

Dear Sir

I was a little late in replying to your two kind letters, because I have been very busy lately—I am very happy to see that the study of M. Jacques Inaudi has greatly interested you.

M. Charcot and I have just recently studied a calculator of the visual type, M. Diamandi,[7] and I will be happy to send you our study, which will appear in five or six days.

I also intend to send you our studies on the electric pen, which is somewhat similar, it seems to me, to things that you have

---

[7] Jean-Martin Charcot (see above); Périclès Diamandi (1868–?) a calculating prodigy born at Pylaros in Greece.

published lately.[8]

Finally, I am currently correcting, for the *Revue des Deux Mondes*, an article on the memory of chess players in which I was able to confirm in the most unexpected way one of the observations you made in your investigation of "Mental Images". I'm not telling you anything about it yet, to leave you in suspense.[9]

Please accept, dear Sir, my best regards,

Alfred Binet

[8] Edison's Electric Pen.

[9] Alfred Binet (1893). "Les grandes mémoires: Résumé d'une enquête sur les joueurs d'échecs". In: *Revue des Deux Mondes* 117 (June 15, 1893), pp. 826–860.

Paris, 29 rue Madame
22 mai 1893

Cher Monsieur

J'ai un peu tardé à répondre à vos deux aimables lettres, parce que j'ai été très occupé dans ces derniers temps – je suis très heureux de voir que l'étude de M Jacques Inaudi vous a vivement intéressé.

Nous venons d'étudier récemment, M Charcot et moi, un calculateur du type visuel, M Diamandi, et je me ferai un plaisir de vous envoyer notre étude, qui va paraître dans cinq ou six jours.

Je compte aussi vous envoyer nos études sur la plume électrique, qui ressemblent un peu, à ce qu'il me semble, à des choses que vous avez fait paraître dernièrement –

Enfin, je corrige en ce moment, pour la Revue des Deux Mondes, un article sur la mémoire des joueurs d'échecs dans lequel j'ai pu confirmer de la manière la plus inattendue une des observations que vous avez faite dans votre enquête sur "Mental Imaging". Je ne vous en dis rien encore, pour vous laisser la surprise.

Veuillez agréer, cher Monsieur, mes meilleurs sentiments,

Alfred Binet

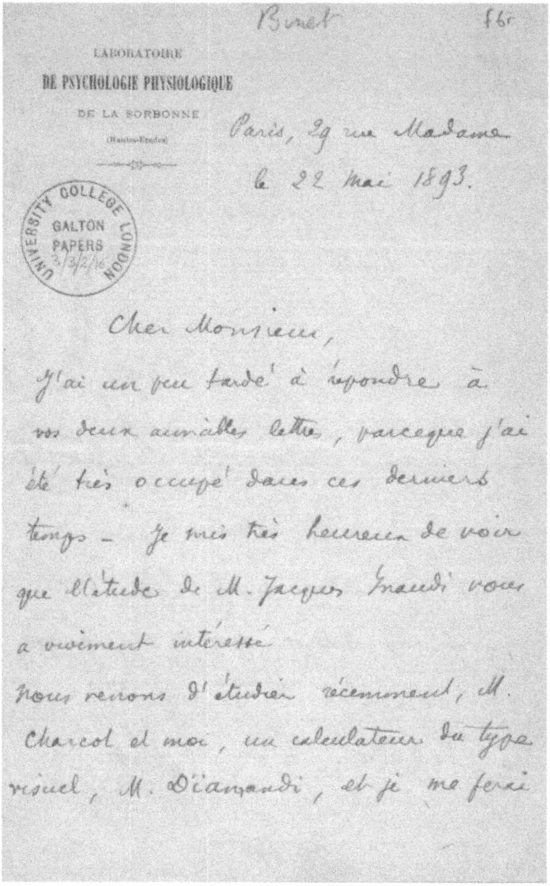

LABORATOIRE
DE PSYCHOLOGIE PHYSIOLOGIQUE
DE LA SORBONNE
(Hautes-Études)

Paris, 29 rue Madame
le 22 mai 1893.

Cher Monsieur,

J'ai un peu tardé à répondre à vos deux aimables lettres, parceque j'ai été très occupé dans ces derniers temps — Je suis très heureux de voir que l'étude de M. Jacques Inaudi vous a vivement intéressé.

Nous venons d'étudier récemment, M. Charcot et moi, un calculateur du type visuel, M. Diamandi, et je me ferai

Figure 2.10: Binet to Galton 1893-05-22 Page 1

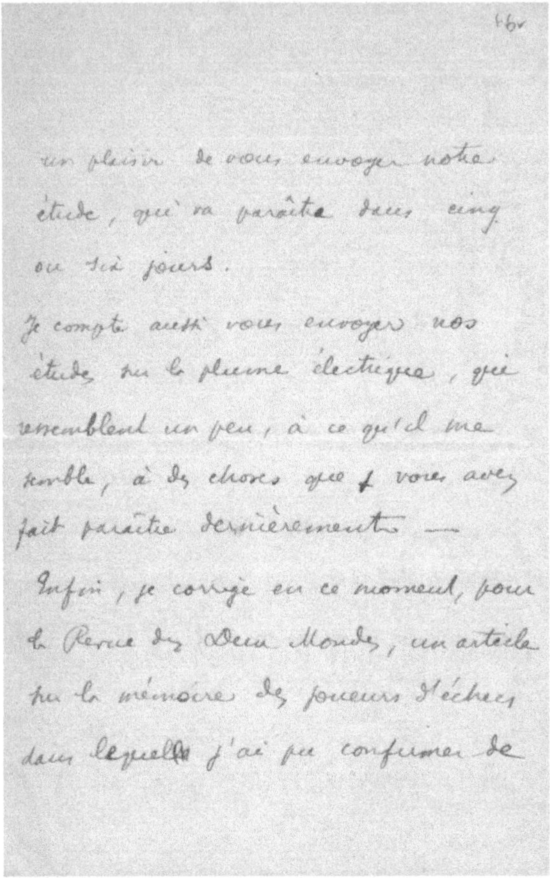
un plaisir de vous envoyer notre
étude, qui va paraître dans cinq
ou six jours.
Je compte aussi vous envoyer nos
études sur la plume électrique, qui
ressemblent un peu, à ce qu'il me
semble, à des choses que vous avez
fait paraître dernièrement —
Enfin, je corrige en ce moment, pour
la Revue des Deux Mondes, un article
sur la mémoire des joueurs d'échecs
dans lequel j'ai pu confirmer de

Figure 2.11: Binet to Galton 1893-05-22 Page 2

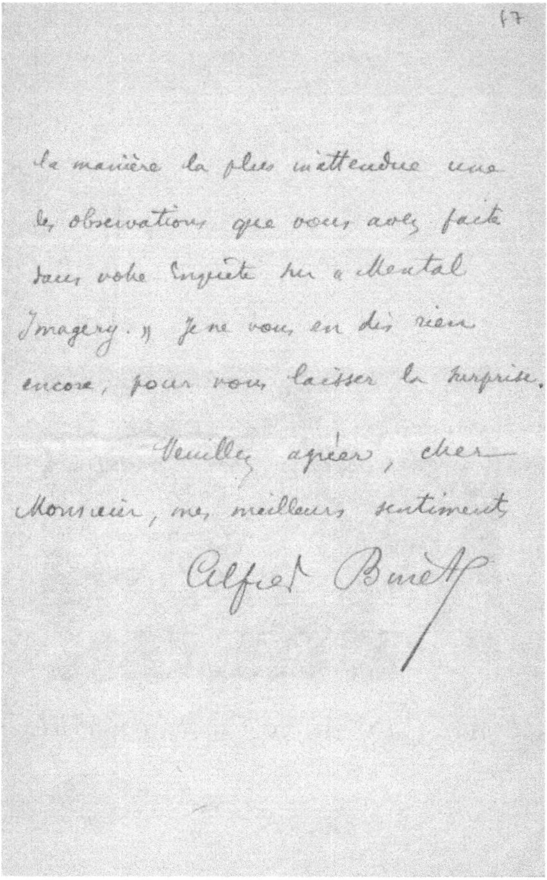
17

de manière la plus inattendue une
les observations que vous avez faites
dans votre Enquête sur « Mental
Imagery. » Je ne vous en dis rien
encore, pour vous laisser la surprise.

Veuillez agréer, cher
Monsieur, mes meilleurs sentiments

Alfred Binet

Figure 2.12: Binet to Galton 1893-05-22 Page 3

Figure 2.13: Alfred Binet in mid-life.

2.7 1894-08-22: Alfred Binet to Francis Galton

Paris, 29 rue Madame
22 August 94

Dear Sir

We are to publish from next year a Psychological Annual summarizing all the works of experimental psychology published in 1894.[10] We would like to devote a small bibliographic notice of 3 to 5 pages to Mr. Romanes, recently deceased, and we ask if you can write this notice.[11] We would be happy to have a contribution from your pen in our Annual.[12]

---

[10] *L'Année Psychologique*, founded by Binet in 1895.

[11] George John Romanes (1848–1894), a Scottish biologist born in Canada but raised and schooled in Scotland from the age of a few months, then at Cambridge. Galton and Charles Darwin corresponded extensively with him.

[12] Galton does not appear to have written any such 'notice'.

Please accept, dear Sir, my most best regards,

Alfred Binet

Paris, 29 rue Madame

22 aout 94

Cher Monsieur

Nous devons publier à partir de l'année prochaine une Année Psychologique résumant tous les travaux de psychologie expérimentale parus en 1894. Nous désirons consacrer une petite notice bibliographique de 3 à 5 pages à M Romanes, récemment décédé, et nous vous demandons si vous voudriez bien vous charger de cette notice. Nous serions heureux d'avoir dans notre Année une contribution de votre plume.

Veuillez agréer, cher Monsieur, mes sentiments les plus distingués,

Alfred Binet

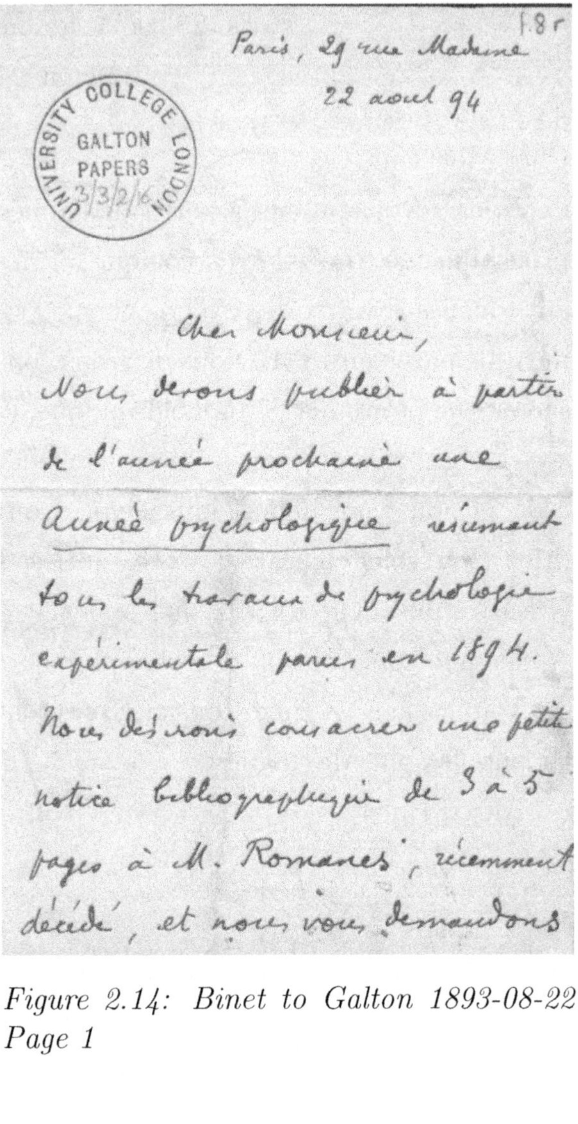

f.8r

Paris, 29 rue Madame
22 août 94

Cher Monsieur,

Nous devons publier à partir de l'année prochaine une Année psychologique résumant tous les travaux de psychologie expérimentale parus en 1894. Nous désirons consacrer une petite notice bibliographique de 3 à 5 pages à M. Romanes, récemment décédé, et nous vous demandons

Figure 2.14: Binet to Galton 1893-08-22 Page 1

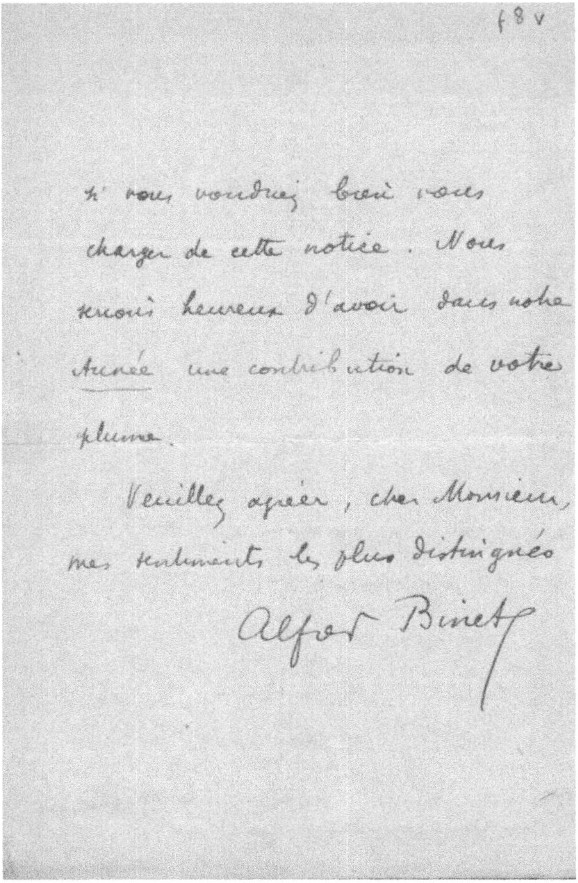
f8v

si vous voudriez bien vous
charger de cette notice. Nous
serions heureux d'avoir dans notre
Année une contribution de votre
plume.

Veuillez agréer, cher Monsieur,
mes sentiments les plus distingués

Alfred Binet

Figure 2.15: Binet to Galton 1893-08-22 Page 2

2.8 1894-10-11: Alfred Binet to Francis Galton

Postcard:

Mister Fr. Galton

42 Rutland Gate

London

England

[11 Oct 1894]

Dear Sir

Would you be kind enough to send me a reprint of your work which appeared in Nature on "Relative Sensitivity of Men and Women",[13] so that I can analyze it in our *Psychological Annual*?

yours faithfully

A. Binet

29 rue Madame, Paris

[13] Francis Galton (1894c). "The relative sensitivity of men and women at the nape of the neck, by Weber's test". In: *Nature* 50 (May 10, 1894), pp. 40–42.

Carte postale:

Monsieur Fr Galton

42 Rutland Gate

Londres

Angleterre

[11 oct 1894]

Cher Monsieur

Seriez-vous assez aimable pour m'envoyer un tirage à part de votre travail paru dans Nature sur "Relative Sensitivity of Men and Women", afin que je puisse l'analyser dans notre Année Psychologique?

Votre bien dévoué

A. Binet

29 rue Madame, Paris

UNIVERSITY COLLEGE LONDON
GALTON PAPERS
CARTE POSTALE
Ce côté est exclusivement réservé à l'adresse
Monsieur Fr. Galton
42 Rutland Gate
Londres
Angleterre
Binet

Figure 2.16: Binet to Galton 1894-10-11 Recto

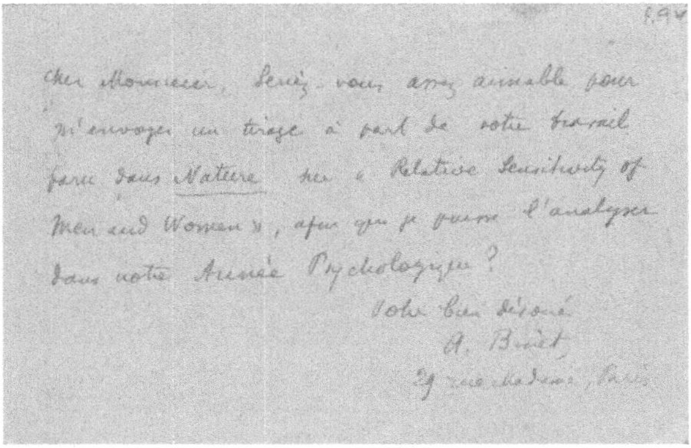
Cher Monsieur, Seriez vous assez aimable pour m'envoyer un tirage à part de votre travail paru dans Nature sur « Relative Sensitivity of Men and Women », afin que je puisse l'analyser dans notre Année Psychologique?
Votre bien dévoué
A. Binet,
[illegible] rue Madame, Paris

Figure 2.17: Binet to Galton 1894-10-11 Verso

Figure 2.18: Composite photo of seven criminals. Binet shared Galton's enthusiasm for composite photographs.

2.9 1894-11-24: Alfred Binet to Francis Galton

Postcard:

Mister Fr. Galton

42 Rutland gate

London

Dear Sir

I have just received from you Nature and I am very happy to read the very kind — too kind — analysis that you present of my latest book.[14] Many thanks. How I regret not seeing you more often! We have just finished researching the recall of ideas which will certainly interest you.

yours faithfully

Alfred Binet

29 rue Madame

[14] Francis Galton (1894b). "Psychology of mental arithmeticians and blindfold chess-players". In: *Nature* 51 (Nov. 22, 1894). Review of Alfred Binet, Psychologie des Grands Calculateurs et Joueurs, pp. 73–4.

24 Nov. 1894

Carte postale:
Monsieur Fr Galton
42 Rutland gate
London

Cher Monsieur

Je reçois à l'instant de votre part Nature et je suis très heureux de lire la si aimable – la trop aimable – analyse que vous présentez de mon dernier livre. Mille remerciements. Comme je regrette de ne pas vous voir plus souvent! Nous venons de terminer des recherches sur la mémoire des idées qui certainement vous intéressent.

Votre bien dévoué
Alfred Binet
29 rue Madame

24 nov. 1894

f.10r

UNIVERSITY COLLEGE LONDON
GALTON PAPERS
3/3/2/16

CARTE POSTALE

Ce côté est exclusivement réservé à l'adresse

Binet

Monsieur Francis Galton
42 Rutland Gate

London

Figure 2.19: Binet to Galton 1894-11-24 Recto

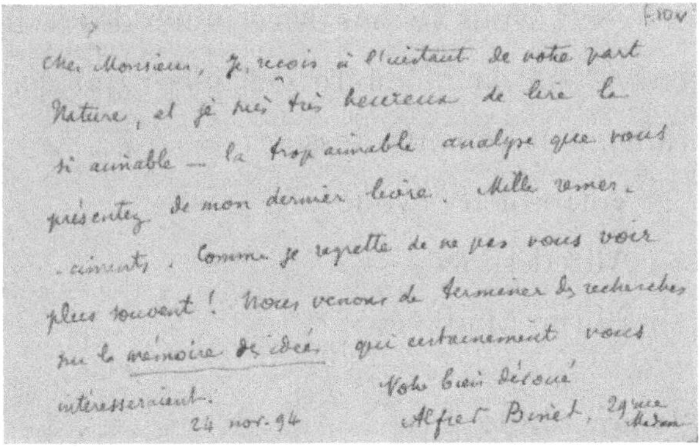
f.10v

Cher Monsieur, Je reçois à l'instant de votre part
Nature, et je suis très heureux de lire la
si aimable — la trop aimable analyse que vous
présentez de mon dernier livre. Mille remer-
-ciements. Comme je regrette de ne pas vous voir
plus souvent! Nous venons de terminer des recherches
sur la mémoire des idées qui certainement vous
intéresseraient.

Votre bien dévoué
24 nov. 94 Alfred Binet, 24 rue Madame

Figure 2.20: Binet to Galton 1894-11-24 Verso

Figure 2.21: Jacques Inaudi.

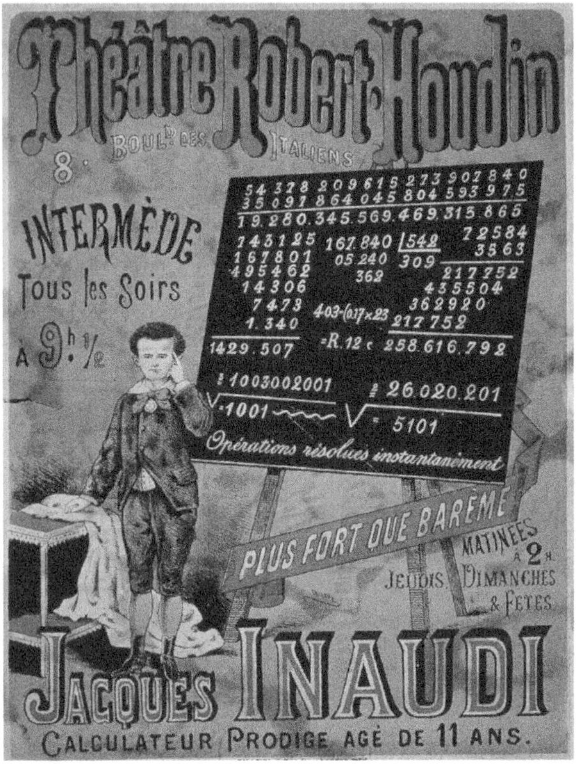

Figure 2.22: Jacques Inaudi appearing at the Theàtre Robert Houdin in 1878. He was then 11 years old.

Figure 2.23: Pericles Diamandi (1868-1932), calculating prodigy. His sister Uranie had similar powers.

Chapter 3

Review by Galton of Binet's *Psychology of Mental Arithmeticians and Blindfold Chess-Players.*

PSYCHOLOGY OF MENTAL ARITHMETICIANS AND BLINDFOLD CHESS PLAY-

ERS.[1]

Psychologie des Grands Calculateurs et Jouers d' Èchecs. Par Alfred Binet. (Paris : Hachette and Cie., 1894.)

Whoever may hereafter write about mental imagery will be imperfectly equipped for his task unless he has mastered the contents of this curious and instructive volume. It analyses the mental processes of two groups of remarkable men — those who possess extraordinary powers of mental arithmetic, and those who are capable of playing eight or more games of chess, blindfold and simultaneously. The idea of making the inquiry is due to the late Prof. Charcot; its prosecution has been conducted almost wholly by M. Binet, and principally at his laboratory in the Sorbonne. The

[1] Francis Galton (1894b). "Psychology of mental arithmeticians and blindfold chess-players". In: *Nature* 51 (Nov. 22, 1894). Review of Alfred Binet, Psychologie des Grands Calculateurs et Joueurs, pp. 73–4.

prosecution of such an inquiry with the accuracy needed by modern psychology is exceedingly difficult, and it is also very difficult to express such results as may be obtained from it, in unambiguous language. The author has, however, succeeded in the latter as well as in the former, and he has framed many happy turns of expression which will contribute to the much desired evolution of psychological language.

The book begins by quoting the series of historical cases of mental arithmeticians, that was published by Scripture in 1881, in the *American Journal of Psychology.* They suffice for making useful generalisations, though few of the cases were tested with much precision. Then the original work commences. It refers to two remarkable calculators, who are now living, both of about the age of twenty-six, but whose mental processes entirely differ in their most obvious characteristics. The one is Inaudi, a Piedmontese, who performs his mental sums wholly, or almost wholly, by imagined sounds, *one*, *two*, *three*, &c.; the other is Diamandi, a Greek, who attains the same end almost wholly by imagined figures, as 1, 2, 3, &c. The careful testing of these two men, and the analyses and comparisons of the results, show the strange unlikeness of human minds in the above well-marked features, accompanied, it may be, with a nearer likeness in those deeper and more obscure qualities, which are exceedingly difficult to grasp. I, myself, had the pleasure of testing Inaudi at my own house, in company with a few scientific friends. Even the small number of experiments that there was then time to make, rendered it clear to my own mind that the conclusions which had been arrived at, after prolonged and careful experiments in France, were quite justified, namely, that he performs his long sums almost wholly by his auditive imagination, supplemented possibly by the motive, or gesture sense, but that the visual form of imagination was practically absent during the calculations. His case is an extremely rare one, and proportionately valuable for study. On the other hand, Diamandi is an excellent example of the common type of mental calculators, who work almost wholly by the visual imagination. A comparison between the achievements of Inaudi and Diamandi under similar tests, is the main feature of the first half of Binet's volume. He succeeds in distinctly negativing the assertion that the visual memory, even of a man who is so exceptionally gifted in that way as Diamandi, resembles actual vision either in its accuracy or in its completeness. Thus if a small square table of twenty-five figures, five figures in breadth and five in height, is shown to and learnt by Diamandi, he takes only nine seconds to repeat them in successive lines, but if he is asked to repeat them in the order of the columns, he is just four times as long in doing so, whether the columns are mentally read from their tops downwards, or from their bottoms upwards. He does not therefore read the figures as if they were written on a mental blackboard, which could be done as easily in any one direction as in any other, but he has, in some obscure way, to puzzle the figures out. When another table of twenty-five figures is taken, in which the figures are variously coloured, Diamandi's power of re-presenting colours being about as strong as that of re-presenting form, he has no difficulty in learning them, but he does it by two successive operations, first learning the figures and then the colours, and he is consequently twice as long over his task. This could hardly be the case if the visualised schedule had the completeness of an "after-image" or of a photographic plate.

A great difficulty in the way of testing the power of the memory of professional calculators is caused by their habit of accumulating large stores of mnemonic helps, which produce results that simulate those of a direct memory. It is indeed difficult for anyone to free himself wholly from the use of such helps, which arise unbidden, more or less consciously, certain runs of figures, or accidents of position in the page, being more readily fixed in the memory than others. Binet's chapter on this subject is very instructive. The most famous calculating boys had their calculating faculties developed very early in life. Many began to calculate of their own accord before they could read or write, and for the most part they were born in humble circumstances. It is found, so far as present information goes, that they did not inherit their gifts, except in a few cases, of which the Bidder family is a conspicuous instance. For my own part, I hesitate for a while to accept the above negative result as a fact, and on the following grounds. Two mental peculiarities have to concur in the making of a calculating boy; the one is a special capacity for mental calculation, and the other is a passion to exercise it. Both of these peculiarities are rare, and they are not necessarily coordinated, therefore the chance of their concurrence in full force may be very small indeed. I have, however, reason to suppose that the capacity for mental calculation is more common than is usually believed, but that it does not commonly interest its possessor, and may even be unknown and consequently neglected by him. Trustworthy evidence for or against its hereditary transmission could hardly be obtained under these conditions. I may quote the case of a deceased lady of remarkable ability, which I indirectly verified to my own satisfaction at the time. She told me, and her husband confirms my recollection, that one night, while travelling to the south of France, she could not sleep, so she amused herself, as is common on such occasions, with various idle trains of thought. Then it occurred to her to try mental sums, and finding, much to her surprise, that she had great facility in doing them, she became interested and exerted herself to the utmost. Before her train had reached Lyons, she had successfully multiplied one series of eleven figures into another series also of eleven figures. She subsequently trained herself to multiply fifteen figures into another fifteen. I am informed that her first attempt at the latter had one error, and, on being told that it was not correct, she went over it again mentally and gave the correct result. Another case comes to my memory. It appears that there was a craze for mental arithmetic in the period 1820-30, or thereabouts. My father was interested in the subject and made experiments on many friends and on all his servants, with the result, as I used to hear, that the best performer of all, and a really remarkable one, was a somewhat obtuse and uninteresting servant girl. She took no especial pleasure in calculation, and on that account would never have made a study of its processes by herself; nevertheless, she had the capacity for using them. An innate passion for arithmetic, such as all the great calculators possessed, is certainly uncommon. If only a moderate passion for it should exist, it is likely to become repressed by circumstances, because it is nearly useless to the possessor. It is difficult to imagine that anyone who was not fascinated by figures would devote the best part of his time and energy to them. Professional calculators are said to be usually (by no means always) narrow-minded, and to have their heads filled with mnemonic contrivances.

I may be permitted to allude to an inquiry analogous to that which has here been made into the visual and auditive imaginations, which I made on myself, on a small scale, in respect to the olfactile imagination. I tried to perform mental arithmetic, not by imaginary visual symbols, or by imaginary sounds, but by imaginary smells. As sums are set in the two former cases, either in really visible symbols or in really audible sounds, while the results are reached through imaginary ones, so in my experiment the sums were set in real odours, and were worked out through imaginary odours. I described the result briefly, not many months ago, in the *American Psychological Review*,[2] and think the inquiry worth repetition, especially by ex-

[2] Francis Galton (1894a). "Arithmetic by smell". In: *Psychological Review* 1 (1894), pp. 61–2.

perimenters who may possess the power of re-presenting odours to themselves more vividly than I have. It would enable them to perceive the processes gone through in mental arithmetic from a new point of view. My apparatus consisted of glass tubes, each drawn to a nozzle at one end like a short syringe. One end of a piece of india-rubber tube, six or eight inches long, was pushed tightly over the other end of the glass. A different odorous substance, camphor, carbolic acid, gasolin, &c., was inserted and packed lightly with cotton wool in the several tubes, whose ends were afterwards tied up. On grasping one of these tubes tightly, at the moment when its nozzle was brought to the nostril, a whiff of its peculiar odour was ejected and simultaneously sniffed up. This could be rapidly repeated three or four times without much diminution of the odour of the whiff. (An arrangement with valves would have much improved its action, by ensuring that no air should be ejected that had not passed through the scent.) I thus possessed a set of tubes that could be used *smellingly*, in the same way as the symbols 1, 2, 3, &c., are used visually, or the words *one, two, three*, &c., are used audibly. This is not the place to enter into further details. I only desire to emphasise one fact which the experiment taught me, namely, the existence of a large substratum of mental work that my power of introspection failed to penetrate. I progressed far enough to be able to add or subtract small sums, so that a 1 followed by a 2, both in smell language, associated themselves at once with the imaginary sniff of a 3, whenever I was engaged in addition, or with that of a 1 when I was engaged in subtraction. But the two associations of 3 and 1 never clashed; they were mutually exclusive. I could not ascertain through introspection what was the nature of the *at-*

titude of mind which determined whether the association was to be the one needed for addition or for subtraction, for division or for multiplication. Another point that strongly impressed me was the enormous amount of labour that must have been gone through by all of us in thoroughly learning the multiplication table. I made a very few similar experiments with the gustatile or taste-imagination, but they were troublesome, and I did not follow them up.

There is little room now left to speak of the latter half of Binet's volume, which refers to the great chess-players, who play eight or more games blindfold and simultaneously. The evidence is overwhelming that the faculty of visualising is not exercised by them in the same sharp and distinct way that it is commonly supposed to be. They do not see the chessmen and the complete board all at once and with clear definition, but they commonly see all besides the portion they axe considering, more or less vaguely, and they appreciate the positions of the men as hidden centres of forces. Two letters, which close the volume, by the distinguished chess-players Goetz and Tarrasch, seem to me models of exact introspection and of clear description.[3]

Francis Galton.

---

[3] Alphonse Goetz (1865–1934) and Siegbert Tarrasch (1862–1934).

Chapter 4

Galton's Notes on Inaudi

May 11 / 93 Thursday.

Signor Ercole[1] had called yesterday with a letter from Prof: Binet to introduce Inaudi. I fixed for a *seance* in my house at 10 am & in the evening at the Royal Society soirée invited a few friends who came.

President Sir Rawson Rawson

Prof. G. J. Romanes[2]

[1] Ercole, C. M., see page 123.
[2] Romanes, George John. See page 124.

Prof: Oliver Lodge

F. Howard Collins

& myself

Also Ercole, the person, a Frenchman, who proposed making a show of Inaudi, and Inaudi himself. Assurances of the most stringent kind were freely given that none of our names should be advertised.

Ercole had given me a translation of Charcot's report to the Institut on Inaudi's history & system of calculating, also the "Evening News & Post", April 1893 which contained in p. 2 a notice of him. [See Figure 4.1, p. 87,]

My first object was to find out how far gesture memory was combined with the auditory memory. To do this it was intended, after the party had become familiarised together, to compare the period required to calculate (1) in his ordinary way with (2) having his mouth open & larynx quiet.

I had — Kreil's tables for multiplication of every 3 figures into every 3 figures; Barlow's tables of squares, cubes &c.; a stop watch.

We began with the multiplication & after a few trials settled down to the plan of my writing out the two triplets of figures, which Ercole read out in French, then Inaudi repeated them to show he had got them. After this he tapped with a pencil to show that he began to calculate. When the calculation was ended he tapped again & forthwith gave the result. I noted the interval by the stop watch between the two taps. ...

Our conclusions were

1. that he was a very straightforward & careful youth (at 25).

2. that he had a prodigious memory for figures.

3. that his calculating powers did not

exceed those of a many bank clerks, who would do the multiplying just as fast as he, if the time occupied in the mechanical operation of writing were subtracted from the whole time they take.

4. He had that wearied look in the eyes & brow that I have seen in other boy calculators.

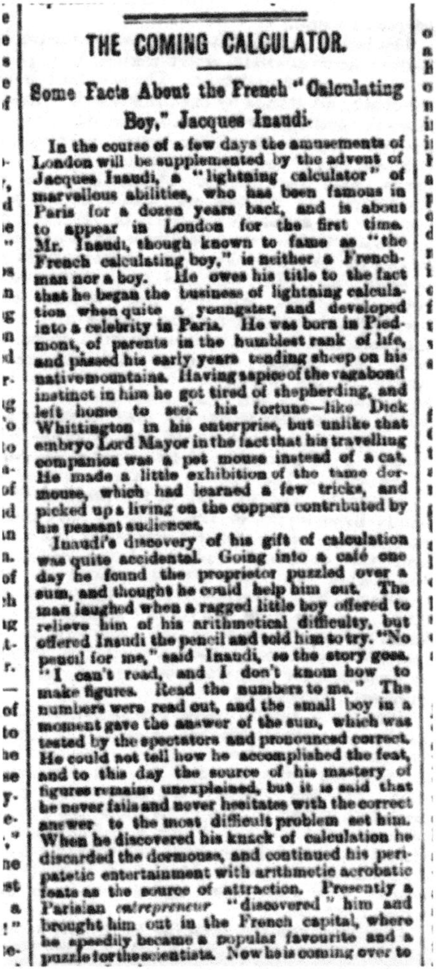

THE COMING CALCULATOR.

Some Facts About the French "Calculating Boy," Jacques Inaudi.

In the course of a few days the amusements of London will be supplemented by the advent of Jacques Inaudi, a "lightning calculator" of marvellous abilities, who has been famous in Paris for a dozen years back, and is about to appear in London for the first time. Mr. Inaudi, though known to fame as "the French calculating boy," is neither a Frenchman nor a boy. He owes his title to the fact that he began the business of lightning calculation when quite a youngster, and developed into a celebrity in Paris. He was born in Piedmont, of parents in the humblest rank of life, and passed his early years tending sheep on his native mountains. Having a spice of the vagabond instinct in him he got tired of shepherding, and left home to seek his fortune—like Dick Whittington in his enterprise, but unlike that embryo Lord Mayor in the fact that his travelling companion was a pet mouse instead of a cat. He made a little exhibition of the tame dormouse, which had learned a few tricks, and picked up a living on the coppers contributed by his peasant audiences.

Inaudi's discovery of his gift of calculation was quite accidental. Going into a café one day he found the proprietor puzzled over a sum, and thought he could help him out. The man laughed when a ragged little boy offered to relieve him of his arithmetical difficulty, but offered Inaudi the pencil and told him to try. "No pencil for me," said Inaudi, so the story goes. "I can't read, and I don't know how to make figures. Read the numbers to me." The numbers were read out, and the small boy in a moment gave the answer of the sum, which was tested by the spectators and pronounced correct. He could not tell how he accomplished the feat, and to this day the source of his mastery of figures remains unexplained, but it is said that he never fails and never hesitates with the correct answer to the most difficult problem set him. When he discovered his knack of calculation he discarded the dormouse, and continued his peripatetic entertainment with arithmetic acrobatic feats as the source of attraction. Presently a Parisian *entrepreneur* "discovered" him and brought him out in the French capital, where he speedily became a popular favourite and a puzzle for the scientists. Now he is coming over to

Figure 4.1: Evening News, 1893-04-28, 2 pt. 1.

perplex London.

The lightning calculator's apparatus for public performances consists of a blackboard, a piece of chalk, and an assistant. The assistant holds the chalk and stands by the blackboard. Inaudi faces the audience with his back to the board, and does not look at the figures as they are written down by his assistant. But he is ready with the answer almost as soon as the figures are written down, and as he proclaims it the assistant records it in its place on the board, so that its correctness may be ascertained by any onlooker with elementary ideas of arithmetic. He will add and subtract long rows of figures, multiply and subtract and divide puzzling combinations, square and cube numbers, and extract square and cube roots, all with the same rapidity and accuracy. These things, however, he regards as the elements of his art. As a specimen of his more ornamental work we may instance the following problem submitted to him at the Sorbonne: Find a number of four figures, the sum of which is 25, the sum of the numbers in place of hundreds and of that in the place of thousands being equal to the figure in ten's place, and the sum of the figures occupying the ten's and the thousand's places being equal to the figure in the unit's place; if the number be reversed it is increased by 8,082. Inaudi gave the correct answer to this brain-racker, the mere recital of which would be enough to give most men a headache, though it is exceedingly easy when understood, in three minutes. It is 1,789.

Sometimes Inaudi does two different sums at the same time. He has done the following two problems simultaneously: To find the cube of a given number, and to tell on what day of the week a date 70 years back fell, giving at the same time the number of hours and minutes which a person born on that date would have lived up to the time of calculation. In person M. Inaudi is small and sturdy, with a big head and a very high and broad forehead.

Londoners are indebted for M. Inaudi's visit

Figure 4.2: Evening News, 1893-04-28, 2 pt. 2.

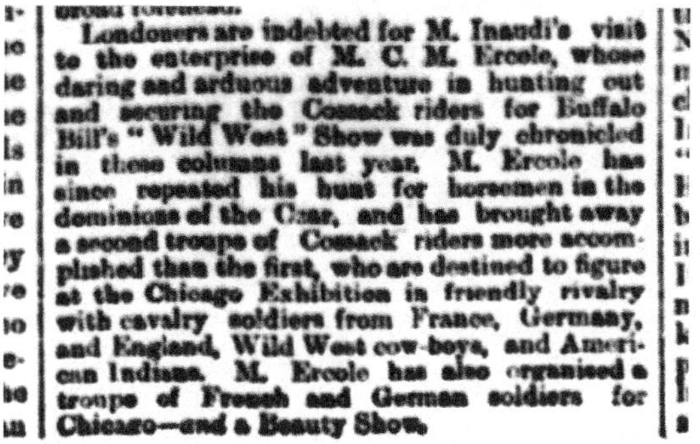

Londoners are indebted for M. Inaudi's visit to the enterprise of M. C. M. Ercole, whose daring and arduous adventure in hunting out and securing the Cossack riders for Buffalo Bill's "Wild West" Show was duly chronicled in these columns last year. M. Ercole has since repeated his hunt for horsemen in the dominions of the Czar, and has brought away a second troupe of Cossack riders more accomplished than the first, who are destined to figure at the Chicago Exhibition in friendly rivalry with cavalry soldiers from France, Germany, and England, Wild West cow-boys, and American Indians. M. Ercole has also organised a troupe of French and German soldiers for Chicago—and a Beauty Show.

Figure 4.3: Evening News, 1893-04-28, 2 pt. 3.

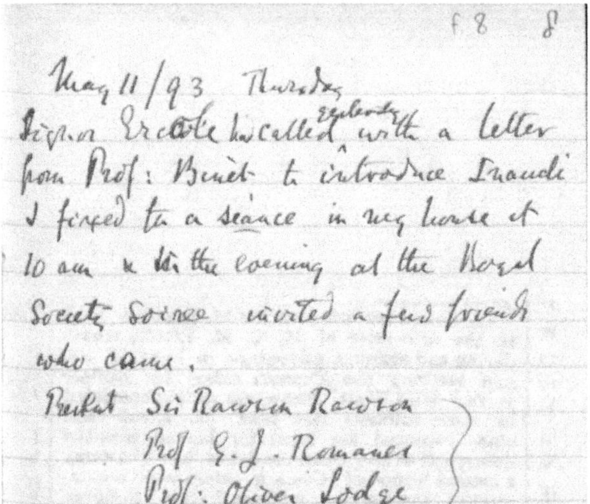
f 8 8

May 11/93 Thursday
Signor Ercole has called yesterday with a letter
from Prof: Binet to introduce Inaudi
I agreed to a séance in my house at
10 am & in the evening at the Royal
Society Soirée invited a few friends
who came.
Present Sir Rawson Rawson
Prof G. J. Romanes
Prof: Oliver Lodge
F. Howard Collins
& myself
Also Ercole, the person, a Frenchman, who professes
making a show of Inaudi, and Inaudi himself
Assurances of the most stringent kind were
freely given that none of our names should
be advertized
Ercole had given me a translation of
Charcot's report to the Institut on Inaudi

Figure 4.4: Galton's notes on Inaudi 01.

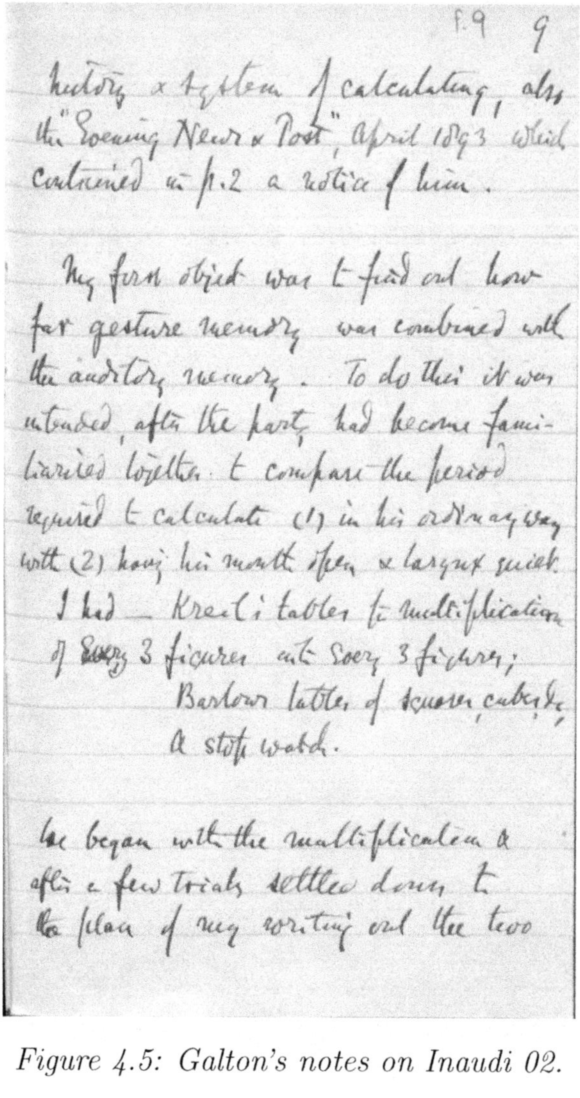

f.9 9

history & system of calculating, also the "Evening News & Post", April 1893 which contained on p.2 a notice of him.

My first object was to find out how far gesture memory was combined with the auditory memory. To do this it was intended, after the party had become familiarised together to compare the period required to calculate (1) in his ordinary way with (2) having his mouth open & larynx quiet.

I had — Krelle's tables for multiplication of every 3 figures into every 3 figures;
Barlow's tables of squares, cubes &c,
a stop watch.

We began with the multiplication & after a few trials settled down to the plan of my writing out the two

Figure 4.5: Galton's notes on Inaudi 02.

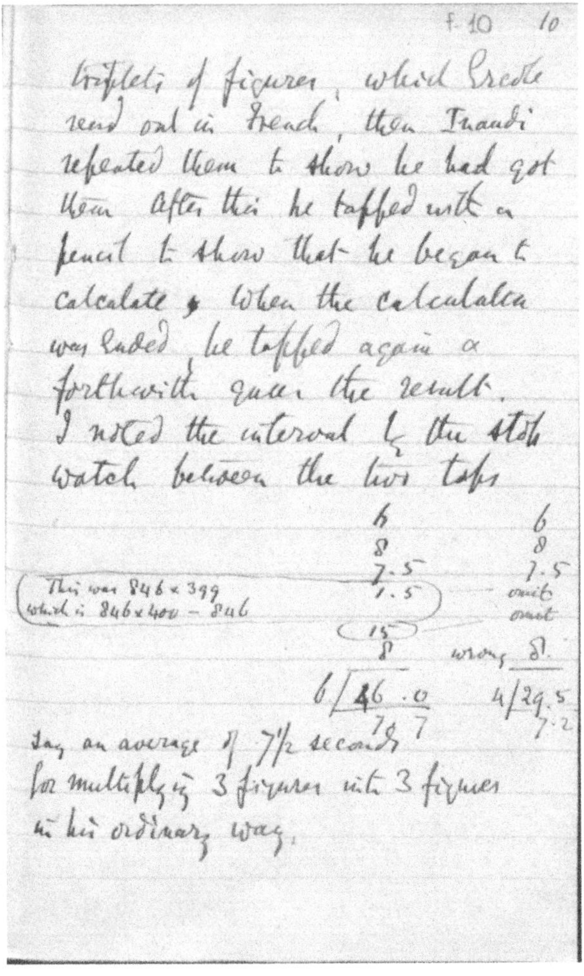
f.10 10

triplets of figures, which [illegible] read out in French, then Inaudi repeated them to show he had got them. After this he tapped with a pencil to show that he began to calculate & when the calculation was ended he tapped again & forthwith gave the result. I noted the interval by the stop watch between the two taps

	6	6
	8	8
	7.5	7.5
This was 846 × 399 which is 846 × 400 − 846	7.5	omit
	15	omit
	8	wrong 8
	6/46.0	4/29.5
	7.7	7.2

say an average of 7½ seconds for multiplying 3 figures with 3 figures in his ordinary way.

Figure 4.6: Galton's notes on Inaudi 03.

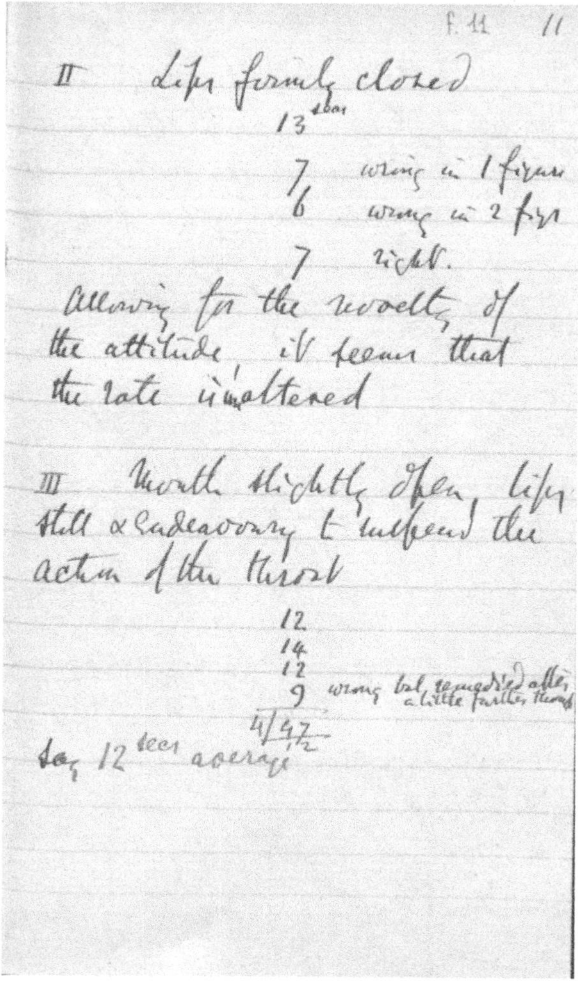
f. 11 11

II Lips firmly closed

13 secs

7 wrong in 1 figure

6 wrong in 2 figs

7 right.

Allowing for the novelty of the attitude it seems that the rate is unaltered

III Mouth slightly open, lips still & endeavouring to suspend the action of the throat

12

14

12

9 wrong but remedied after a little further thought

4/47

12

say 12 secs average

Figure 4.7: Galton's notes on Inaudi 04.

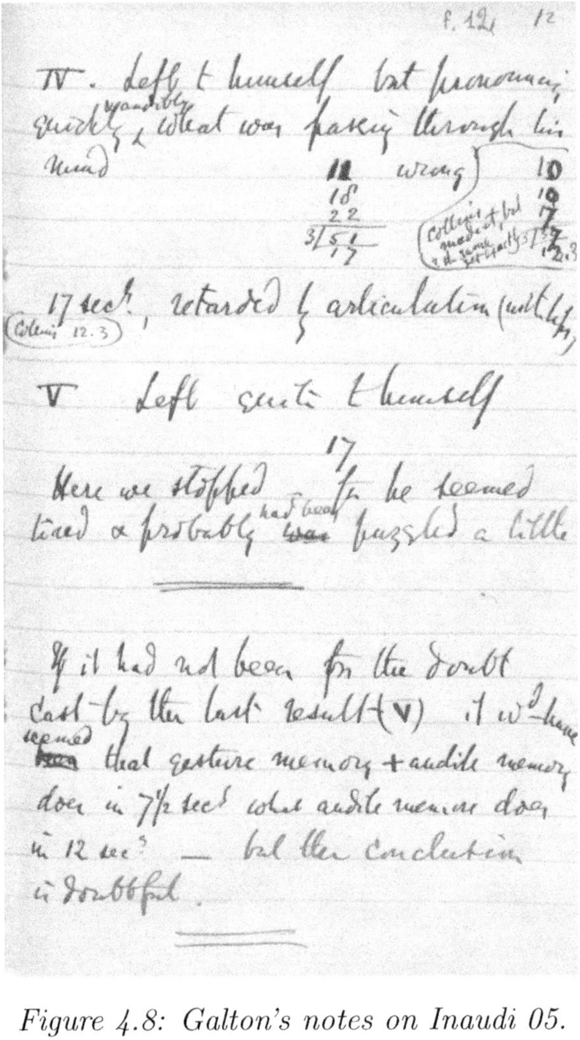

Figure 4.8: Galton's notes on Inaudi 05.

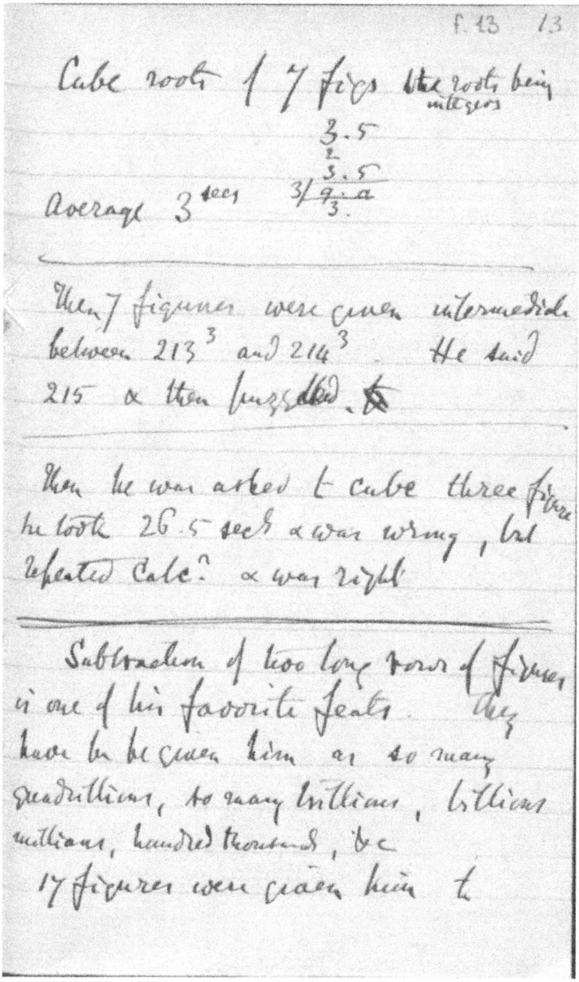
f. 13 13

Cube roots of 7 figs the roots being integers

3.5
2
3.5
3)9.0
3.

Average 3 secs

Then 7 figures were given intermediate between 213^3 and 214^3. He said 215 & then puzzled.

Then he was asked to cube three figures he took 26.5 secs & was wrong, but repeated Calc.n & was right

Subtraction of two long rows of figures is one of his favorite feats. They have to be given him as so many quadrillions, so many trillions, billions millions, hundred thousands, &c

17 figures were given him to

Figure 4.9: Galton's notes on Inaudi 06.

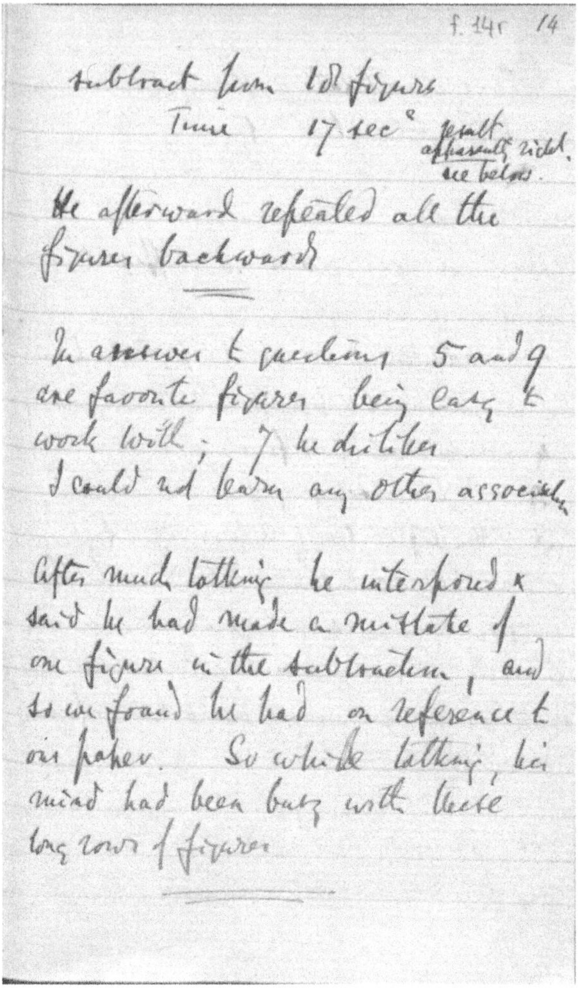

f. 145 14

subtract from 101 figures
Time 17 sec^s ~~result~~ apparently right. see below.

He afterward repeated all the figures backwards

In answer to questions 5 and 9 are favorite figures being easy to work with; 7 he dislikes
I could not learn any other associations

After much talking he interposed & said he had made a mistake of one figure in the subtraction, and so we found he had on reference to our paper. So while talking, his mind had been busy with these long rows of figures

Figure 4.10: Galton's notes on Inaudi 07.

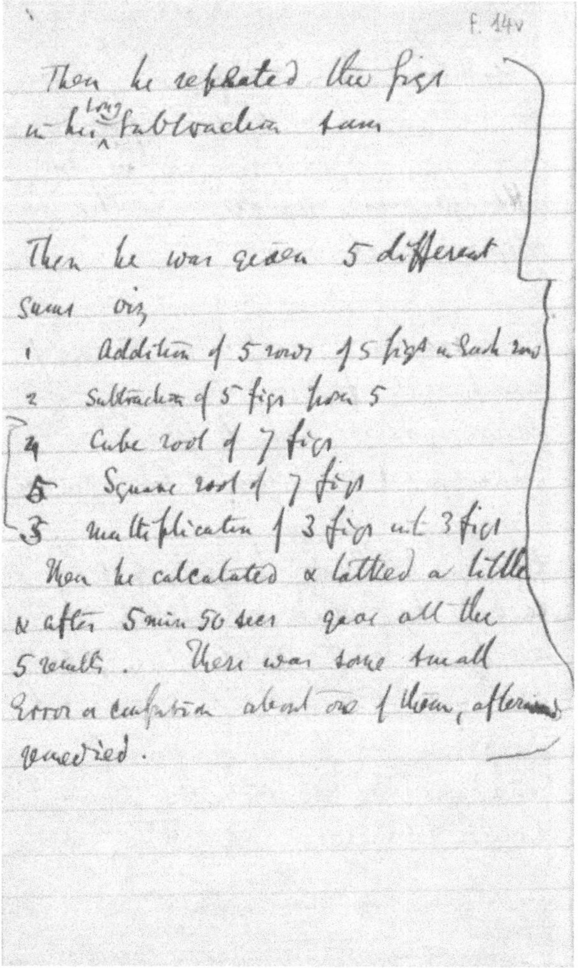
F. 14v

Then he repeated the figs
in his long subtraction sum

Then he was given 5 different
sums viz
1 Addition of 5 rows of 5 figs in each row
2 Subtraction of 5 figs from 5
4 Cube root of 7 figs
5 Square root of 7 figs
3 multiplication of 3 figs with 3 figs
Then he calculated & talked a little
& after 5 min 50 secs gave all the
5 results. There was some small
error or confusion about one of them, afterwards
remedied.

Figure 4.11: Galton's notes on Inaudi 08.

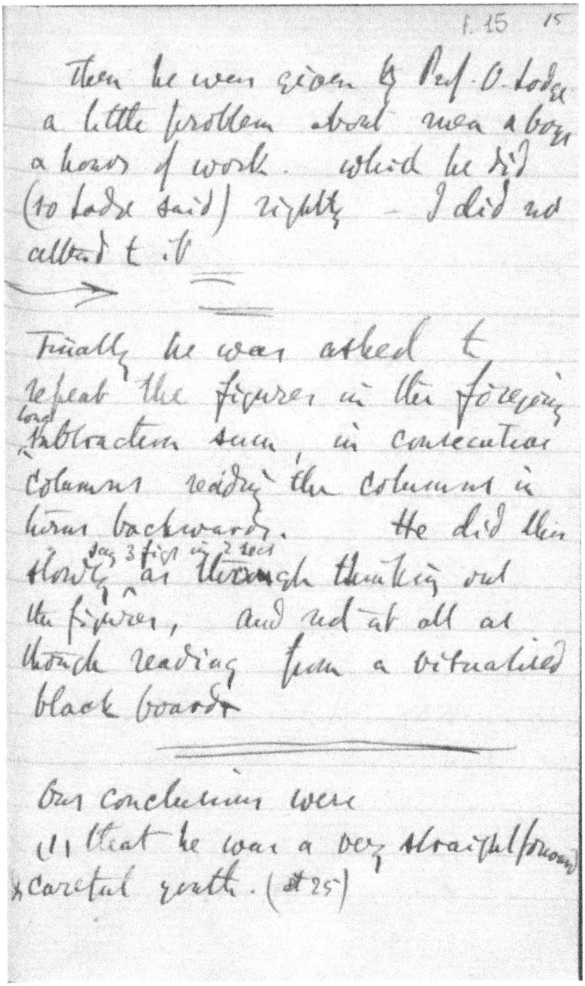

f. 15 15

Then he was given by Prof. O. Lodge a little problem about men & boys & hours of work. which he did (so Lodge said) rightly — I did not attend to it

Finally he was asked to repeat the figures in the foregoing long subtraction sum, in consecutive columns reading the columns in turns backwards. He did this slowly (say 3 figs in 2 secs) as through thinking out the figures, and not at all as though reading from a visualised black board

Our conclusions were
(1) that he was a very straightforward & careful youth. (at 25)

Figure 4.12: Galton's notes on Inaudi 09.

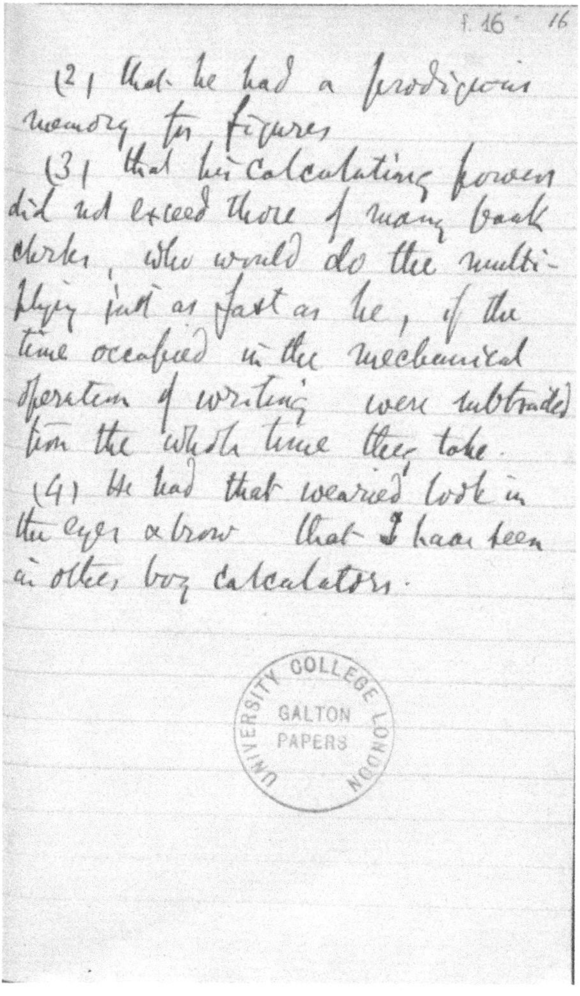
f. 16 16

(2) that he had a prodigious memory for figures

(3) that his calculating powers did not exceed those of many bank clerks, who would do the multiplying just as fast as he, if the time occupied in the mechanical operation of writing were subtracted from the whole time they take.

(4) He had that wearied look in the eyes & brow that I have seen in other boy calculators.

Figure 4.13: Galton's notes on Inaudi 10.

Chapter 5

Charcot's Report on Inaudi

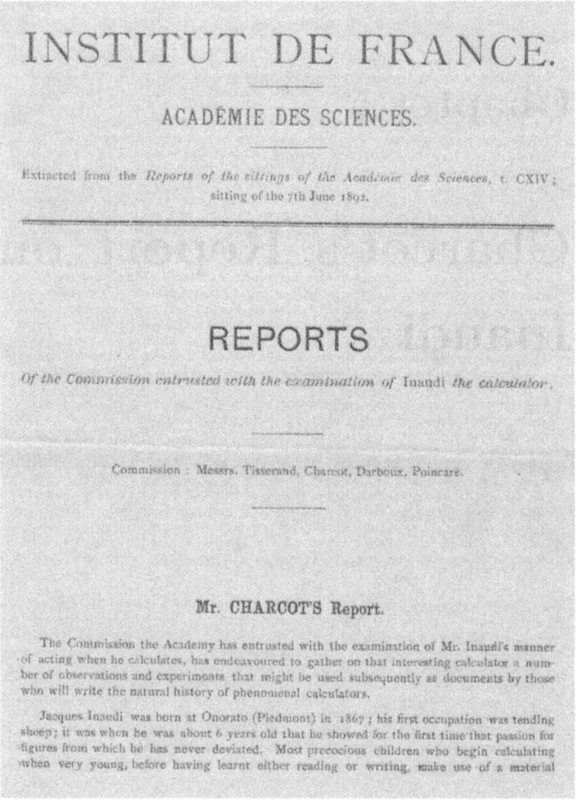

INSTITUT DE FRANCE.

ACADÉMIE DES SCIENCES.

Extracted from the *Reports of the sittings of the Académie des Sciences*, t. CXIV; sitting of the 7th June 1892.

REPORTS

Of the Commission entrusted with the examination of Inaudi *the calculator*.

Commission : Messrs. Tisserand, Charcot, Darboux, Poincaré.

Mr. CHARCOT'S Report.

The Commission the Academy has entrusted with the examination of Mr. Inaudi's manner of acting when he calculates, has endeavoured to gather on that interesting calculator a number of observations and experiments that might be used subsequently as documents by those who will write the natural history of phenomenal calculators.

Jacques Inaudi was born at Onorato (Piedmont) in 1867; his first occupation was tending sheep; it was when he was about 6 years old that he showed for the first time that passion for figures from which he has never deviated. Most precocious children who begin calculating when very young, before having learnt either reading or writing, make use of a material

Figure 5.1: Charcot et al on Inaudi, 1.

— 2 —

numeration—they count on their fingers or with pebbles. Young Inaudi did not picture to himself in that manner the numbers which he combined in his head; he merely had recourse to the names of the numbers which his brother had taught him, reciting them before him. This curious circumstance may have had on Mr. Inaudi's manner of calculating an influence which we shall mention hereafter. Thanks to his prodigious aptitudes, the young shepherd improved rapidly. He soon left his country to follow his parents in their progress through the south of France, collecting coppers and making a marmotte dance; wishing to increase his trifling profits, he offered passers by to solve some problems mentally, and he made his appearance in several public establishments.

In 1880, being then 12 years old, he came to Paris, and was presented to the *Société d'Anthropologie* by Broca, who, after having analysed his manner of acting when reckoning, added: "He can neither read nor write; he has the figures in his head, but he does not write them." Ever since, under the influence of constant exercise, he has widened the sphere of his operations; at twenty he learned to read and to write; though his tardy instruction has remained rudimentary on a very great number of points, his apprehension is very good and wit sharp; he is naturally mild and modest. He is to day a young man of 24, short (1 m. 52), looking robust, normally built; his cranium, actually plagio cephalic, presents in front a slight projection of the right frontal bump, and behind, a projection of the left parietal bump; at the back of the interparietal suture, a longitudinal crest may be felt, 0,02, formed by the turned up right parietal bone; his ears are symetrical and detached from the head like funnels; his face is slightly asymetrical, the right side being smaller than the left, the facial angle is almost a right one (89°); the other cranio-facial mensurations do not show any remarkable anomaly. The methodical examination of his sight and hearing did not lead to the discovery of any alteration or hypercuity in those organs.

The Commission has endeavoured with the greatest care to demonstrate the psychological aptitudes that allow Mr. Inaudi to resolve complex problems by a mere mental operation, namely, without the assistance of either reading or writing. It is beyond all dispute that memory must perform, in those circumstances, the principal part; without constituting, properly speaking, the faculty of reckoning, the known quantities of the problem and its partial solutions could not be recollected without it, up to the moment when the definitive solution is found. Consequently the Commission have thought that they must first of all examine the state of Mr. Inaudi's memory.

The anatomico-clinical investigations of these last years have contributed to demonstrate that the faculty of the mind vulgarly called *memory* is nothing but a complexus, an ensemble. In conclusion, there are but partial, special, or, as they also say, local memories, enjoying a relative reciprocal independance; and *if*, in the conditions that may be called normal, the respective development of the various forms of memory go, as it were, on an equality, it was to be foreseen that in certain anormal conditions, one of them might become weaker, or on the contrary, acquire an excessive development without the necessary participation of the others. It is so in the reality of things, and it is not absolutely exceptional, for instance, that the activity of one of the memories should acquire considerable proportions separately, and even sometimes reach a degree so very much above the common measure, that it excites both astonishment and admiration.

Figure 5.2: Charcot et al on Inaudi, 2.

— 3 —

In the category of these extraordinary partial memories, the hypermnesy of figures and numbers occupies, in some measure the first rank; it is that hypermnesy, at the least, which, perhaps on account of the conditions of appreciation seemingly simple in which it presents itself, attracts most particularly the attention of observers. Mr. Inaudi is a remarkable example of it.

The ensemble of interrogatories and experiments to which he has been submitted has well shown that with him, the memory of colours, forms, events, places, musical airs, etc., does not exceed the normal average, and even remains inferior to the average; he cannot picture to himself the pieces of a chess-board and its squares; he wonders how people can conceive a game mentally; he does not seem endowed with any exceptional aptitude, outside of figures and numbers, for which he shows a remarkable memory. He often dreams of figures, numbers and calculations, and sometimes resolves, in those cases, problems of which he has not found the solution in the course of the day; they are the only dreams of which he keeps, when awake, a distinct recollection, whereas the dreams that only refer to the ordinary events of life leave but a slight impression.

In order to appreciate exactly the extent of the memory of figures with Mr. Inaudi, it may be compared with another memory, that of letters and words. When requested to repeat a certain number of letters or words that have just been pronounced in his presence, Inaudi shows himself incapable of reproducing more than five or six; and under the same circumstances, he cannot recollect, after a first hearing, two lines of either prose or poetry. Whereas he can, without any fatigue, without any hesitation, and with an absolute precision, repeat long series of figures, varying, for instance, from 25 to 30, which he has only heard expressed once. He reproduces the series at will either as it has been recited or inversely; he can even, when asked to do so, keep the recollection of the figures for several weeks. At the end of a sitting, during which numerous problems had been proposed to him, Mr. Inaudi could repeat without any mistake, all the figures and in the order in which the problems had been stated; the number of these figures amounted to *two hundred and thirty-two;* at another meeting he could repeat as many as four hundred.

The extent, precision and suppleness of that special memory of the figures have given rise to a great number of experiments the particulars of which would be too numerous, and which have well demonstrated that, with regard to memory, Inaudi is not second to any of the calculating prodigies who preceded him. One single instance will suffice to give an idea of it. Cauchy sets forth, in his interesting report, the following experiment to which the Commissioners had submitted *Mondeux*, the calculator. To learn a number of twenty-four figures divided into four sets, so as to be enabled to enounce at will, the six figures contained in every one of them, it took Mondeux five minutes. Now, Inaudi has learned a number of twenty-four figures divided into similar sets; he has repeated the second and third set, then the first set backwards, and in the last place the whole number, beginning with the last figure, the whole in fifty-nine seconds.

Another question also relating to the memory of figures has then attracted the attention of the Commission. The question was to know what is the nature of the mnemonic images that Inaudi has recourse to in order to picture to himself the numbers of his operations. The

Figure 5.3: Charcot et al on Inaudi, 3.

— 4 —

investigations concerning that psychological process have led to an important observation, which we think must modify the current notions about calculating prodigies.

If we consult the few biographical studies that have been published up to this day on the most celebrated calculators, and which are found recorded in a recent article of Mr. Scripture's [*Arithmetical Prodigies* (*Americ. Journ. of Psych.*, April, 1891)], we find that these calculators use, according to their own testimony, as a principal basis of their mental operations, the visual memory. At the very moment the known quantities of the problem are enounced before them, they give themselves the internal vision of the numbers enounced, and these numbers remain, for the length of time requisite for the operation, before their imagination as if they were written on a fictitious board placed before their eyes. This process of *visualisation* was Mondeux's, Colborn's, of all those, in one word, who have had an opportunity of explaining themselves clearly.

On that subject Bidder, another wonderful calculator, has even written in his memoirs that he would not understand the possibility of mental calculation without the faculty of representing to one's self the figures as if they were seen. Besides, it seems to appear from Mr. Galton's investigations, that many calculators operate on visual images in which the figures are sometimes written on lines or grouped in forms, the shapes of which vary with the individuals.

The study of Inaudi's process shows that a general conclusion could not be drawn from the preceding facts. Though it may seem rational to admit that the simplest means, for a calculator, in order to replace the black board and written figure, which he does not see, is to give himself a visual representation of board and figure, it must be admitted that it is possible to attain the same result by processes of a quite different nature. Inaudi has not recourse to mental vision, but to mental hearing. His testimony, the attitude he assumes while reckoning, and the different tests which he stood remove every doubt with regard to this question. When questioned by the Commission on his subjective impressions, he answered without hesitation: "I hear the numbers, and it is the ear that retains them. While I am trying to reproduce them from memory, I hear them resound inwardly, with the tone of my own voice, and I continue hearing them for a good part of the day. After one hour or two hours, if I wish to think of a number that has been enounced, I can repeat it as exactly as I have done just now."

Subsequently the Commission returning to this important question, Inaudi developed his first assertion with much lucidity and intelligence: "I do not see the figures, said he, I must even say that I find it more difficult to recollect the numbers and figures when they are communicated to me in writing, than by word of mouth. I feel very uneasy in the first case. I do not like to write figures myself; writing them would not help me remembering them; I prefer to hear them." On another occasion, Inaudi made the following remark, which it is useful to retain; having begun to learn to read and write four years ago, he could not have pictured to himself before that time written figures, since he did not know them. The Commission has been in a position, several times, to verify the correctness of those assertions. It is certain that Inaudi calculates with more facility when the known quantities of the problem are communicated to him by word of mouth, than when they are written down before him;

Figure 5.4: Charcot et al on Inaudi, 4.

— 5 —

the sight of written figures embarrasses him, and then, coming back to his natural processes, he recites himself either loud or in a low tone of voice, the numbers which he must retain in his memory. It must also be remarked that when a series of numbers is enounced before him, he must necessarily articulate them loud to fix and preserve them in his memory, and while he is fixing them, just the same as when he is reckoning, he may be heard muttering very rapidly the names of the figures. This articulation of the numbers forms an integral part of his reckoning processes, so much so that any contrivances interfering with that articulation abates the rapidity of the calculation and renders it less correct.

A direct experiment, the result of which is not uninteresting, has allowed the testimony of the matter on these delicate questions to be controlled.

After having disposed on a sheet of paper, in squares, five numbers of five figures each, the chess-board was presented to Mr. Inaudi, begging him to learn it by heart. He did so conformably to his usual method, namely, reading the numbers audibly. Then he was requested to enounce by heart either diagonally or following such or such vertical or horizontal line of the chess-board. He succeeded, not without difficulty, after many hesitations. If Inaudi belonged to the category of sightsmen, he would be obliged to grope, and he would read the answer before him, without any hesitation, the same as on an imaginary board.

The conclusion to be retained is that Inaudi, differently from several of the calculators who preceded him, has not recourse to visual memory for his mental operation ; he avails himself concurrently of auditive images and motive images of articulation. Which of these two elements is predominant ? Is it the motive or sensitive element ? The absence of an experimental process allowing to separate them, does not allow to determine their respective importance. But the articulation of the figures only seems to intervene in order to strengthen the phenomena of internal hearing, which necessarily have the priority. This is, besides, Mr. Inaudi's own opinion.

The Commission, after having ascertained Mr. Inaudi's characters of precocity and impulse to calculation met with in the history of calculating prodigies, asked themselves under the influence of which anthropological conditions this young calculator has developed himself. It is known that in certain cases, when individuals seemed to be endowed very early with remarkable aptitudes, either similar aptitudes could be found with other members of their families (as for instance in the families of celebrated musicians), or phenomena of nevropathic heredity. The inquiry instituted by the Commission on these important questions has unfortunately only produced results, which, for the greatest part, were negative. Heredity, though investigated carefully, only revealed a few oddities and lacks of co-ordination of character with the paternal ancestor ; it does not appear that Mr. Inaudi's brothers or other persons of his family have ever presented special aptitudes of any description. The previous conduct of the individual is of no consequence, and the anthropological examination which he underwent has only set forth, as one can see, a small number of stigmas of little importance. The Commission express the wish that the attention of observers be called henceforth to these questions, and that all the family conditions capable of throwing light on so considerable and anormal development of certain psychical faculties be collected.

Figure 5.5: Charcot et al on Inaudi, 5.

— 6 —

Mr. DARBOUX'S Report.

To the so very interesting report the Academy has just heard, the Commission have thought that they must add a few details on the manner in which Inaudi executes the arithmetical operations that are submitted to him, and they have entrusted me with this part of the report. The task has been rendered easy for me by the innumerable experiments to which Inaudi readily submitted. He held himself at ourdisposal, and at that of leading scfentific men, and the information we gathered is as complete as we could wish. The result of our examination strikes us as deserving to be communicated to the Academy ; but in order to throw some light on our statement we consider that it is indispensable to separate, in Inaudi, the calculator who effects elementary arithmetical operations, and the man who resolves in a more or less complete manner, the problems of mathematics, the solution of which he is desired to produce.

I shall first speak of the calculator. Let us report it at first, the really extraordinary results which we have witnessed rest, before anything else, on a prodigious memory. At the end of a sitting offered to the pupils of our universities, Inaudi has repeated a series of numbers comprising more than 400 figures, and, if there have been one or two hesitations, he has had recourse to no one (he even begged not to be assisted), to rectify the trifling errors he had committed, or to find again figures a little forgotten.

At one of our meetings we gave Inaudi a number of 22 figures. A week afterwards he could repeat it though we had not told him that we should renew the question. It is not necessary to insist on facts of that description ; but we must add that Inaudi's memory has much improved by exercise. Only a few years ago, at Lyons, he contented himself with multiplying numbers by 3 figures. To-day, he can make multiplications each of the factors of which has at least 6 figures. These operations are made with an extraordinary rapidity, and Inaudi has certainly effected in less than ten seconds, the cube of 27.

A second point, which strikes us as being most interesting, has been passed over by the persons who examined him. The very simple processes Inaudi has recourse to to execute the different operations, have been analysed carefully, but there is a self-evident fact that has not been sufficiently remarked : namely, that these processes have been imagined by the calculator himself,—that *they are quite original*. For instance, while Mondeux and many other prodigies had been instructed by men who acquainted them with the usual methods, Inaudi, never having had a master, has certainly imagined the rules which he applies to every one of his operations. And, what is interesting, is that these rules are different from those which are taught every where in Europe in the elementary schools, where as some resemble in certain respects those which are followed by other nations, by the Hindoos for instance, which will be illustrated by the following statement.

Addition. Inaudi adds up easily 6 numbers of from 4 to 5 figures ; but he proceeds successively, adding the two first, then the sum to the following, and so on. He always begins his addition by the left, the same as *the Hindoos do to-day*, instead of beginning by the right, as we do.

Figure 5.6: Charcot et al on Inaudi, 6.

— 7 —

Subtraction. It is one of Inaudi's triumphs. He subtracts easily one from the other two numbers of about 20 figures, *again beginning by the left.*

Multiplication. The processes which he follows are quite elementary, but they require Inaudi's memory. For instance, in order to multiply 834×36, he makes the following decompositions:

$$\left.\begin{array}{r} 800 \times 30 = 24\,000 \\ 800 \times 6 = 4\,800 \\ 30 \times 36 = 1\,080 \\ 4 \times 36 = 144 \end{array}\right\} \text{Sum Total: } 30\,024$$

In all those partial multiplications, one of the factors never has more than one significant figure. Nevertheless he knows and uses the property of the factor 25; he knows that, in order to multiply by that number, it suffices to take the fourth of the hundred-fold. For instance, for the square of 27, he will make the following decomposition.

$$\left.\begin{array}{r} 25 \times 27 = 675 \\ 2 \times 27 = 54 \end{array}\right\} \text{Sum Total} \ldots\ 729$$

Sometimes he uses partial products with the sign —. For instance, for the cube of 27, viz., the product of 729 by 27, he will effect the decomposition:

$$\left.\begin{array}{r} 700 \times 20 \\ 700 \times 7 \\ 30 \times 20 \\ 30 \times 7 \end{array}\right\} \begin{array}{rr} \text{or } 730 \times 27 = & 19\,710 \\ -27 & \underline{27} \\ & 19\,683 \end{array}$$

Division. Here, Inaudi follows the ordinary rule, that brings division back to a subtraction; but sometimes, with the simplification his memory allows him to have recourse to, a memory to which we must always come back.

Powers. For the production of squares, Inaudi knows and applies the relative to the square of a sum. For instance, for the square of 234 567, he uses the decomposition:

$$\overline{234\,000}^2 + 2 \times 234\,000 \times 567 + \overline{567}^2.$$

Extraction of roots. Here no rule is followed; there is nothing but mere groping. For instance, to find a root which is 14 672, Inaudi will have tried 14 000 and 15 000, then 14 500 and 14 600, then 14 650, 14 660, 14 670, and every time the power of the number tried will have been deducted from the given number.

Figure 5.7: Charcot et al on Inaudi, 7.

— 8 —

For roots of a superior order, it is evident that the operation is so much the easier as the indicator of the root is more elevated. This is what those do not always understand who are amazed at the extraction of a fifth root.

We must say a few more words of the problems which Inaudi has begun to solve by himself, these last years. We do not speak here of the questions that are obviously brought back to a series of calculations. Inaudi has, for instance, been able to name rapidly the total number of corn grains, which, they say, the inventor of the game of chess claimed as a reward. He merely calculated and added successively the numbers of grains that were to be placed on every one of the squares of the chess-board. But he has been able sometimes to solve the most difficult questions of arithmetic and algebra, the solution of which was given by integer numbers. He could find rapidly entire roots of certain algebric equations; but when we gave him problems leading to equations of the first degree, we found that his processes were mere groping, and that he began by supposing that the solutions sought for were integer. It can hardly be otherwise. He cannot be expected to find again by himself the whole of algebra and mathematics. But we have ascertained that he is intelligent, and of a sound mind. If we also remark that the memory with which he is endowed has been met with in several celebrated mathematicians, we must regret that at the age when he could study, he did not receive the lessons of an intelligent and clever master.

Imprimerie SYMONDS, 6, rue Ordener et rue Lallier, 10, Paris.

Figure 5.8: Charcot et al on Inaudi, 8.

Figure 5.9: Alfred Binet with daughters Madeleine and Alice.

Figure 5.10: Alfred Binet with subjects.

Figure 5.11: Alfred Binet in mid-life.

Figure 5.12: Alfred Binet in late life.

Figure 5.13: Alfred Binet with colleagues Jules Courtier and Jean Philippe, and a student, Victor Henri, at the Sorbonne in 1893.

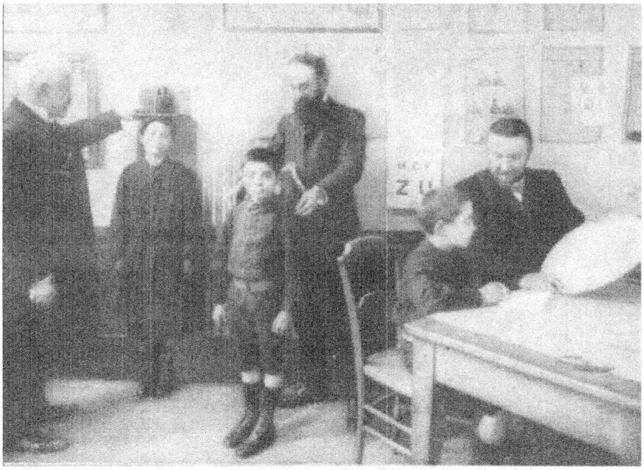

Figure 5.14: Alfred Binet with schoolchildren.

Figure 5.15: Alfred Binet with kymograph.

Figure 5.16: Theodore Simon.

PUBLICATIONS OF THE TRAINING SCHOOL AT VINELAND
NEW JERSEY

DEPARTMENT OF RESEARCH

THE INTELLIGENCE OF THE FEEBLE-MINDED

BY

ALFRED BINET, Sc.D. AND TH. SIMON, M.D.

TRANSLATED BY

ELIZABETH S. KITE

Diplome d'Instruction Primaire Superieure
Paris le 23 Juillet, 1906
Member of the Staff of the Vineland
Research Laboratory

THE TRAINING SCHOOL ★ AT VINELAND, NEW JERSEY ★
DEVOTED TO THE INTERESTS OF THOSE WHOSE MINDS HAVE NOT DEVELOPED NORMALLY

NO. 12, JUNE, 1916

Figure 5.17: Elizabeth Kite's translation of Binet and Simon, 1916.

List of Figures

1.1 Alfred Binet. 9

1.2 Francis Galton in the 1890s. 10

1.3 Jacques (Giacomo) Inaudi as a boy. 11

1.4 Binet with Promoter. . . . 12

1.5 Galton at Rutland Gate. . . 13

1.6 Galton in 1904. 14

1.7 Galton in 1904. 15

1.8 The Binet Family. 16

2.1 Binet to Galton 1891-11-25 Page 1 20

2.2 Binet to Galton 1891-11-25 Page 2 21

2.3 Alfred Binet in profile . . . 22

2.4 Binet to Galton 1892-03-19 Page 1 26
2.5 Binet to Galton 1892-03-19 Page 2 27
2.6 Binet to Galton 1893-02-10 30
2.7 Binet to Galton 1893-02-12 34
2.8 Binet to Galton 1893-04-20 Page 1 38
2.9 Binet to Galton 1893-04-20 Page 2 39
2.10 Binet to Galton 1893-05-22 Page 1 45
2.11 Binet to Galton 1893-05-22 Page 2 46
2.12 Binet to Galton 1893-05-22 Page 3 47
2.13 Alfred Binet in mid-life. . . 48
2.14 Binet to Galton 1893-08-22 Page 1 52
2.15 Binet to Galton 1893-08-22 Page 2 53

2.16 Binet to Galton 1894-10-11 Recto 57
2.17 Binet to Galton 1894-10-11 Verso 57
2.18 Composite criminals. 58
2.19 Binet to Galton 1894-11-24 Recto 62
2.20 Binet to Galton 1894-11-24 Verso 62
2.21 Jacques Inaudi. 63
2.22 Jacques Inaudi Poster. . . . 64
2.23 Pericles Diamandi. 65

4.1 Evening News, 1893-04-28, 2 pt. 1. 87
4.2 Evening News, 1893-04-28, 2 pt. 2. 88
4.3 Evening News, 1893-04-28, 2 pt. 3. 89
4.4 Galton's notes on Inaudi 01. 90
4.5 Galton's notes on Inaudi 02. 91
4.6 Galton's notes on Inaudi 03. 92
4.7 Galton's notes on Inaudi 04. 93

4.8 Galton's notes on Inaudi 05. 94
4.9 Galton's notes on Inaudi 06. 95
4.10 Galton's notes on Inaudi 07. 96
4.11 Galton's notes on Inaudi 08. 97
4.12 Galton's notes on Inaudi 09. 98
4.13 Galton's notes on Inaudi 10. 99

5.1 Charcot et al on Inaudi, 1. . 102
5.2 Charcot et al on Inaudi, 2. . 103
5.3 Charcot et al on Inaudi, 3. . 104
5.4 Charcot et al on Inaudi, 4. . 105
5.5 Charcot et al on Inaudi, 5. . 106
5.6 Charcot et al on Inaudi, 6. . 107
5.7 Charcot et al on Inaudi, 7. . 108
5.8 Charcot et al on Inaudi, 8. . 109
5.9 Alfred Binet with daughters Madeleine and Alice. 110
5.10 Alfred Binet with subjects. 110
5.11 Alfred Binet in mid-life. . . 111
5.12 Alfred Binet in late life. . . 111
5.13 Alfred Binet and Sorbonne colleagues 112
5.14 Binet with schoolchildren. . 113

5.15 Alfred Binet with kymograph. 114
5.16 Theodore Simon. 115
5.17 Binet and Simon. 116

Register

Ercole, C. M.

A Parisian impresario, active in the late 19th century. 'I never knew whether he was French or English. He spoke French perfectly, and English like an Oxford don. He made one fortune and retired; lost it in speculation, and went back to his office in the Chaussée d'Antin', Charles B. Cochrane (1926). *The Secrets of a Showman.* New York: Henry Holt, 1926, 192 83

Romanes, George John

1848–1894. Evolutionary biologist of Scottish descent. Though he was born in Canada, he left when only a few months old. He studied at Cambridge, becoming a FRS in 1879. He corresponded extensively with Charles Darwin and Francis Galton. 83

Bibliography

Binet, Alfred (1886). *La psychologie du raisonnement, recherches expérimentales par l'hypnotisme.* Paris: Alcan, 1886.

— (1893). "Les grandes mémoires: Résumé d'une enquête sur les joueurs d'échecs". In: *Revue des Deux Mondes* 117 (June 15, 1893), pp. 826–860.

Burt, Cyril (1952). "Autobiography". In: *History of Psychology in Autobiography.* Ed. by E. G. Boring. Vol. 4. Clark University Press, 1952.

Cochrane, Charles B. (1926). *The Secrets of a Showman.* New York: Henry Holt, 1926.

Galton, Francis (1887a). "Supplementary notes on 'Prehension in idiots'". In: *Mind* 12 (Jan. 1887), pp. 79–82.

— (1887b). "Thoughts without Words [Letters]". In: *Open Court* 1 (1887), pp. 441–4, 472–4, 498–50.

— (1888). "Remarks on Replies by Teachers to Questions Respecting Mental Fatigue". In: *Journal of the Anthropological Institute* 18 (Apr. 10, 1888), pp. 157–68.

— (1894a). "Arithmetic by smell". In: *Psychological Review* 1 (1894), pp. 61–2.

— (1894b). "Psychology of mental arithmeticians and blindfold chess-players". In: *Nature* 51 (Nov. 22, 1894). Review of Alfred Binet, Psychologie des Grands Calculateurs et Joueurs, pp. 73–4.

— (1894c). "The relative sensitivity of men and women at the nape of the neck, by Weber's test". In: *Nature* 50 (May 10, 1894), pp. 40–42.

Haier, Richard (2017). *The Neuroscience of Intelligence.* London: Cambridge University Press, 2017.

Jensen, Arthur R. (1980). *Bias in Mental Testing.* New York: Free Press, 1980.

— (1998). *The g Factor. The Science of Mental Ability.* Westport, Connecticut: Praeger, 1998.

Klein, Alexandre (2016). "À la recherche de l'archive psychiatrique perdue. L'histoire des fonds d'archives d'Alfred Binet (1857-1911) [Searching for the Lost Psychiatric Archive. History of the Alfred Binet's Archives Collections]". In: *L'archive psychiatrique* 41.2, automne (2016), pp. 21–32.

— ed. (2018). *Archives Familiales (1883–1916).* Vol. 3. Correspondance

D' Alfred Binet. Paris: L'Harmattan, 2018.

Tredoux, Gavan (2023a). *Francis Galton's Genius. 1865–1911*. galton.org, 2023.

— (2023b). *Francis Galton's Nature and Nurture. 1822–1865*. galton.org, 2023.

Made in the USA
Monee, IL
04 November 2026

67385775R00075